THE SLIDE AREA

FILM BOOK REVIEWS
1989-1991

Also by Anthony Slide

Early American Cinema (1970)
The Griffith Actresses (1973)
The Idols of Silence (1976)
The Big V: A History of the Vitagraph Company (1976; revised 1987)
Early Women Directors (1977; revised 1984)
Aspects of American Film History prior to 1920 (1978)
Films on Film History (1979)
The Kindergarten of the Movies: A History of the Fine Arts Company (1980)
The Vaudevillians (1981)
Great Radio Personalities in Historic Photographs (1982)
A Collector's Guide to Movie Memorabilia (1983)
Fifty Classic British Films: 1932-1982 (1985)
A Collector's Guide to TV Memorabilia (1985)
The American Film Industry: A Historical Dictionary (1986)
Great Pretenders (1986)
Fifty Classic French Films: 1912-1982 (1987)
The Cinema and Ireland (1988)
Sourcebook for the Performing Arts (with Patricia King Hanson and Stephen L. Hanson, 1988)
The International Film Industry: A Historical Dictionary (1989)
Silent Portraits (1990)
The Television Industry: A Historical Dictionary (1991)
Nitrate Won't Wait: A History of Film Preservation in the United States (1992)

With Edward Wagenknecht:

The Films of D.W. Griffith (1975)
Fifty Great American Silent Films: 1912-1920 (1980)

As Editor:

Selected Film Criticism: 1896-1911 (1982)
Selected Film Criticism: 1912-1920 (1982)
Selected Film Criticism: 1921-1930 (1982)
Selected Film Criticism: 1931-1940 (1982)
Selected Film Criticism: 1941-1950 (1983)
Selected Film Criticism: Foreign Films 1930-1950 (1984)
Selected Film Criticism: 1951-1960 (1985)
International Film, Radio, and Television Journals (1985)
The Best of "Rob Wagner's Script" (1985)
Selected Theatre Criticism: 1900-1919 (1985)
Selected Theatre Criticism: 1920-1930 (1985)
Selected Theatre Criticism: 1931-1950 (1986)
Filmfront (1986)
Selected Radio and Television Criticism (1987)
Selected Vaudeville Criticism (1988)
The Picture Dancing on a Screen (1988)
They Also Wrote for the Fan Magazines (1992)

THE SLIDE AREA

FILM BOOK REVIEWS
1989-1991

Anthony Slide

British Library Cataloguing-in-Publication data available

Library of Congress Cataloging-in-Publication Data

Slide, Anthony.
The Slide area : film book reviews. 1989-1991 / by Anthony Slide.
p. cm.
Includes indexes.
ISBN 0-8108-2614-3 (acid-free paper)
1. Motion pictures—Book reviews. 2. Television—Book reviews.
I. Title.
PN1995.S535 1992
791.43'75—dc20 92-32388

Manufactured in the United States of America

Printed on acid-free paper

For Sue Laimans

Editor of *Classic Images* during the years
in which these reviews were written

Contents

Introduction

"The Slide Area" is a familiar sign to visitors to the Los Angeles area and is the name of a novel by Gavin Lambert. It is also the title of a monthly book review column, published in *Classic Images*. This volume gathers all the reviews written for that column from its inception in January 1989 through December 1991. The more than 300 reviews are primarily devoted to film books, but also included are all major books on television and a sampling of other books related in one way or another to film. General film books make up the bulk of the reviews, while biographies and autobiographies are given less attention. In writing "The Slide Area" column, I have always tried to review as many current film books as possible and have never purposely ignored a book—no matter how little its subject matter might appeal to me.

My book reviews are, I hope, entertaining, and should provide some general pleasure and enlightenment—and hopefully a fair amount of irritation with my opinions—to the casual reader. In that, on a monthly basis, I have tried to review all relevant film books as they appear, *The Slide Area* should be particularly helpful to librarians in tracking down reviews and bibliographic information on the ever-growing number of film books published in the United States. Not only are reviews of film books from all major publishers included here, together with reviews of all film books from smaller academic publishers (such as, Greenwood Press, McFarland, and Scarecrow), but perhaps more important are the reviews of obscure volumes, often self-published or issued by publishers unfamiliar to most librarians.

These books are generally not reviewed in such standard library reference periodicals as *Choice* and *Library Journal*, and, of course, they are never documented in the pages of *Publishers Weekly*. A further major difference between my reviews and those in the aforementioned and similar journals is that they are all written by the same individual. As someone who has spent more than twenty years in the field of film research and film scholarship, I am extremely familiar with film literature of both the past and present. In fact, personally I have added more than forty volumes to the subject, and I am someone who is often described by reviewers with that cliched adjective, "indefatigable."

In one of the volumes reviewed here, *Film Study*, its author Frank Manchel describes much of the narrative in my books as "terse, passionate and nostalgic." Those are adjectives which might well be used to describe my reviews. A thorough reading of my reviews will also reveal that I am easily angered by unreadable, academic prose. I do not like to be written down to by an author, nor I suspect do most other readers. The necessity for a dictionary permanently at hand does not make for easy, or satisfactory, reading—and books which are impossible to read have little value in this nation's libraries.

My reviews do contain personal commentary and thought, but they also indicate all other similar books on the subject, comparing the current book in question with earlier volumes on the same topic. With my background in the subject, I am able to point out major errors and omissions in texts, which are often ignored or overlooked by general review publications. A reference book which has a scholarly "look" but in which I can find three or four errors to an entry is of little use to a librarian or researcher. Unlike other reviewers who merely skim the surface, I study the texts and look for what is positive and negative in the volumes.

I make no excuse for my reviews being opinionated. A review of a stage play, a motion picture, or a book which does not express the reviewer's personal ideology is not a valid one. At the same time, I have tried to be fair in my reviews. I have not, and never will, based an opinion on my personal like or dislike for an author. I have been harsh at times on professional writers, but that is, to me, only fair, because a professional writer should expect criticism. Where a book is self-published and is, quite obviously, a labor of love, I have been a little, and sometimes a lot, kinder in my approach.

Each review includes a general summary of the book's content. Also included with each review is bibliographic data as to price (at the date of publication), publisher, the number of pages, whether the book is available in hardcover or paperback, and whether it contains an index and/or illustrations. The number of pages given is based on my personal examination of the book. I mention this because I have noticed many publishers list a number of pages which bears no connection to the actual number of pages in the published book.

If I have had a personal connection with a book, generally as its editor, the volume is included here, with general information as to its content, but I offer no opinion as to its worth.

Reviews are reprinted here in chronological order, year by year. A title index and an author index provides speedy access to a particular review. Also included is a list of publishers of cited books, with their addresses. Many of the publishers listed here are obscure, and this is possibly the only source for information as to their location. At the same time, it must be pointed out that one-time publishing entities often disappear quickly, and there is no guarantee that a publisher of a particular book is still operational.

"The Slide Area" continues to be published on a monthly basis by *Classic Images* (P. O. Box 809, Muscahie, Iowa 52761) and readers anxious to keep up-to-date on new film books are invited to subscribe.

Anthony Slide

1989

January 1989

The Suspense Thriller: Films in the Shadow of Alfred Hitchcock by Charles Derry. $35. (McFarland. Hardcover. 351 pages. Index. Illus.)

The Suspense Thriller is what its title suggests, with Part I subtitled "Toward a Definition of the Suspense Thriller," and Part II titled "Different Kinds of Thrillers." The author argues for the suspense thriller as a genre (a quite reasonable request), trying to explain just what constitutes such a film and the genre's historical relevance and future. There is little that is new here, although I did learn what is the meaning of "philobat" (one who enjoys thrills) and an "ocnophil" (one who does not). Whether this new knowledge is worth $35 is questionable.

Disability Drama in Television and Film by Lauri E. Klobas. $39.95 (McFarland. Hardcover. 496 pages. Index.)

With almost 400 entries, arranged under headings such as Blindness, Wheelchair-Users, Deafness, Amputation, Developmental Disability, and Small-Stature, *Disability Drama in Television and Film* is an interesting and certainly unusual offering. Each entry includes credits, a brief plotline, and an evaluation from a disability viewpoint. Certainly, the book is in no way definitive; there are no entries on Lon Chaney's *The Penalty* and *The Unknown* under Amputation, nor on Chaney's *Hunchback of Notre Dame* under Multiple Disabilities (!). In fact, silent films are just about ignored, except for such obvious titles as Chaplin's *City Lights,* while television programs receive considerable attention. Yet, to a certain extent, this makes the book more useful—it is often easier to find documentation on classic silent films than it is on episodes of *The Paper Chase* or *T. J. Hooker.* A worthwhile introduction to an unusual topic.

Memorable Films of the Forties by George Aachen. 2 volumes. $40 each. (Rastar Press. Hardcover. Each volume 256 pages. Illus.)

Published in Australia, and intended for the Australian market, these two volumes provide credits, synopses, and commentary on dozens of features from *Abbott & Costello Meet Frankenstein* to *Zombies on Broadway.* If you want to know when a film was released in Australia and what its Australian running time is, then these books are for you. Otherwise, it is more a matter of whether the writing style—gushing film buff—appeals.

Feminist in the Dark by Kathi Maio. $7.95. (The Crossing Press. Paperback. 223 pages. Illus.)

Usually I cringe at the thought of having to read a volume of criticism by a feminist. Happily, such is not the reaction to *Feminist in the Dark.* Kathi Maio is highly readable, and her comments are thought-provoking. She is absolutely correct in that *Wish You Were Here* is a movie of sexual abuse and, after reading Maio's review, I felt ashamed that I left the theatre amused by the film. *Wish You Were Here* is one of more than thirty contemporary features reviewed here. Recommended.

American Film Studios: An Historical Encyclopedia by Gene Fernett. $35 (McFarland. Hardcover. 295 pages. Index. Illus.)

American Film Studios gathers together Gene Fernett's writings from *Classic Images,* and the book is rightly dedicated to the paper's founding editor, Sam Rubin. There is a lot of useful information here, together with a fair amount of highly opinionated, sometimes flowery writing. More than sixty studios are covered, some in more detail than others, with very little information provided on the recent years of ongoing studios—the entry on Paramount does not go beyond 1954, and there is no mention of M-G-M's lot now being owned by Lorimar.

Framework by Tom Stempel. $22.95. (Continuum. Hardcover. 294 pages. Index. Illus.)

The first history of screenwriting in the American film and a first-rate job by Stempel, who ingenuously begins his book by announcing "Some of what I told you before was wrong." Stempel divides the volume into three sections, "Silent Beginnings," "The Studio Period," and "Independent Production," and provides good documentation and analysis of the work of writers as varied as Gene Gauntier, Frank Woods, Dudley Nichols, and Ben Hecht, many of whom are discussed in relationship to the studios with which they were associated. There is an introduction by Philip Dunne (whom at times I find a little tedious), a chapter on the Writers Guild, and even two pages on screenwriting for animation. Highly Recommended.

Film Review, 1988-9 edited by F. Maurice Speed. (Columbus Books. Hardcover. 192 pages. Index. Illus.)

For British children interested in film, one of the delights of the holiday season was, and is, a gift of F. Maurice Speed's *Film Review,* now in its 44th year. Each volume contains a survey of films of the year, with credits and brief synopses; a list of TV features; obituaries; a lengthy book review section by Ivan Butler; and Speed's inimitable commentary on the year in the cinema. It may be too late to try and acquire the previous 43 issues, but there is no excuse for not starting to collect *Film Review* from its current issue. The volume has no U.S. publisher, but is available in the States from specialty film bookshops.

The Dead That Walk by Leslie Halliwell. $24.95. (Continuum. Hardcover. 262 pages. Index. Illus.)

Leslie Halliwell is best known for his useful reference work, *Halliwell's Filmgoer's Companion,* now in its umpteenth edition. *The*

Dead That Walk is a highly readable examination of some of the more popular films featuring Dracula, Frankenstein, mummies, and zombies. It is not intended for the serious scholar in that it has no bibliography and no indication as to sources, apart from an acknowledgement to "the many previous writers on allied subjects." Films are discussed in great detail as regards plots, and a considerable portion of what Halliwell claims is the original script for *Bride of Frankenstein* is reprinted. What is perhaps most admirable about Halliwell's work is that he does show a good knowledge of literary sources, far more so than many other writers in the horror genre.

Documents in Canadian Film edited by Douglas Featherling. $16.95. (Broadview Press. Paperback. 343 pages.)

Documents in Canadian Film is nothing more nor less than a compilation of items relating to Canadian cinema, beginning with a review of the Cinématograph from an 1896 edition of the Montreal newspaper, *La Presse,* and ending with a 1982 essay by James Dunn on "The Death of Canadian Cinema." Intended as a sister volume to *Documents in Canadian Art,* published in 1987, *Documents in Canadian Film* is precisely the type of volume for which there is an urgent need. It is unfortunate that there are not more such volumes reprinting source materials on American or British cinema.

February 1989

Treasures from the Film Archives: A Catalog of Short Silent Fiction Films Held by FIAF Archives edited by Ronald S. Magliozzi. $62.50. (Scarecrow Press. Hardcover. 834 pages. Indexes.)

Here is a superb catalog documenting the holdings—with information as to title, production company, director, and actors—of 33 member archives of the International Federation of Film Archives. (All the U. S. archives have contributed to the book.) Indexes in each category further enhance the value of a volume which includes films from 25 countries. However, there is a depressing element to the book in that Eileen Bowser's preface, together with a "Note to Users," makes it very plain that none of these archives will guarantee that it will allow anyone to view any of the films for whatever purpose. This negativism, so typical of some archivists, raises a fundamental question as to the purpose of film preservation. Are films to be held captive in archives, or are they to be free for study and reevaluation. One can only suggest that future U. S. government funding for film preservation be subject to a commitment by archives that their holdings are accessible to the public. Despite my reservations as to the workings of some of our film archives, I have no problem in praising *Treasures from the Film Archives.* Highly Recommended.

March 1989

The American Film Institute Catalog: Feature Films, 1911-1920 edited by Patricia King Hanson. $125. (University of California Press. Hardcover. 2 volumes of 1,081 pages and 476 pages. Indexes.)

Without question, the most important publishing event in the area of film for this year (and probably the decade) is the issuance of *The American Film Institute Catalog: Feature Films, 1911-1920.* The catalog has a new publisher, the University of California Press, taking over from Bowker (which produced the earlier volumes on the Twenties and Sixties), and a number of new features, which are worth noting.

Aside from personal name, corporate name, subject, and literary and dramatic source indexes, the 'teens volume boasts a chronological index of titles, a genre index, a geographic index (indicating where on location various films were shot), and a selected bibliography (which serves as a first-rate reading list for the silent era). There are changes also in the manner in which information on individual features is recorded. Following each entry are the names, dates and page numbers for the various periodicals which provided source documentation on that title. Where modern sources offer additional or differing information that is presented in a Note, which for many films also offers historical background on the production.

It is unfortunate that it has taken some fifteen years for these two volumes to appear from the date when I first started preparatory work for the project back in 1973. However, the books have appeared at last, and the wait has been worthwhile.

Credit for publication belongs to David Packard, who influenced the David and Lucile Packard Foundation to provide major funding for the research and to sponsor the actual publication. It is unfortunate that there are not more David Packards in the world, and unjust that he was not asked to write the Foreword to these volumes, rather than the Chairman of the National Endowment for the Humanities, which came a distant second in funding.

Credit and our thanks also belong to Executive Editor Patricia King Hanson and Assistant Editor Alan Gevinson, without whose dedication these volumes could neither have appeared nor been published with such meticulous attention to detail. Everyone interested in silent films, on whatever level of scholarship or enthusiasm, owes a tremendous debt to Pat and Alan. And all those who care about sound films should be delighted to know that these same two are currently hard at work in preparation of *The American Film Institute Catalog: Feature Films, 1931-1940.*

Twentieth Century-Fox: A Corporate and Financial History by Aubrey Solomon. $29.50. (Scarecrow Press. Hardcover. 285 pages. Index.)

In his introductory note, the author writes, "This volume [No. 20 in the Filmmakers series] is not intended to be a complete history of the studio and its movies. Instead, it covers a general overview of the finances (in laymen's terms, not accountants') and corporate practices in producing and releasing the studio's output. As well, it tries to present a clear development of the movie business and audience trends, charting a path to the crazy quilt of financing and producing patterns in today's market."

The Cinema of the Sea: A Critical Survey and Filmography, 1925-1986 by Tony Thomas. $24.95. (McFarland. Hardcover. 248 pages. Index. Illus.)

A readable study of ships, sailors, and the sea, with chapters on Pirates, Flags and Glory, Mutiny, Drama on the High Seas, Nautical Fun, War on Water, Beneath the Waves, Disaster, Under Sail, and Salt

Water Gothics. The author has paid scant attention to shipboard romances claiming "love at sea is not intrinsically different from love on land." He does not explain why his survey starts, arbitrarily, in 1925. The filmography is useful, as are Thomas's general, nonfilm-related comments on the subjects to hand.

Interviews with B Science Fiction and Horror Movie Makers by Tom Weaver. $29.95. (McFarland. Hardcover. 413 pages. Index. Illus.)

An excellent collection of interviews with 28 individuals associated with the topic at hand. This is probably the only source for interviews with some of the participants, such as Reginald LeBorg, and, in a way, it is unfortunate that the questions are limited to the interviewee's science fiction and horror films. I particularly enjoyed hearing from Howard W. Koch (not to be confused with the screenwriter), who displays a good, honest attitude towards filmmaking, and who gives credit to people who are overlooked. A good book.

Directed by Jack Arnold by Dana M. Reemes. $24.95. (McFarland. Hardcover. 243 pages. Index. Illus.)

Reemes has produced a good, solid critical study of the career of the director responsible for some 27 features from 1950-1976, the best known of which are *It Came from Outer Space, Creature from the Black Lagoon,* and *The Incredible Shrinking Man.* The one, puzzling aspect of the book is a strange introduction, a "Forryword" by Forrest J. Ackerman, which adds nothing whatsoever to the volume and is written in a very curious style.

Kitty by Kitty Carlisle Hart. $18.95. (Doubleday. Hardcover. 263 pages. Index. Illus.)

Too many of my friends regard Kitty Carlisle as a figure of fun. I enjoy her—what other reason could one have had for watching *To Tell the Truth*?—and I am happy to report that her autobiography is tastefully written, scrupulously accurate, and, my friends will be glad to learn, the lady is not above poking fun at herself. Indeed, Miss

Carlisle even goes so far as to state that she is an excellent straight woman for comedians such as Groucho Marx. Of course, much of the book is taken up with her marriage to Moss Hart, but there are interesting revelations with regard to her relationships with George Gershwin and Sinclair Lewis, and I particularly enjoyed her account of being chased about the bedroom by Bernard Baruch in his long underwear, while she was worrying, "I must get this old gent back under the covers—he'll catch pneumonia!" Great fun—a great book from a great lady.

April 1989

Science Fiction, Horror & Fantasy Film and Television Credits Supplement: Through 1987 compiled by Harris M. Lentz, III. $65. (McFarland. Hardcover. 924 pages. Indexes.)

This is an update to the author's earlier two volumes (published in 1983) and covers films published between then and 1987, together with additional information on some earlier productions. The main body of the text consists of a listing of appropriate films with actor credits, years of production, country of origin if not the United States, and directors. Surprisingly, no other technical credits, not even the writers, are provided, although one of the complementary indexes claims to include directors, producers, writers, artists, etc. In other words, when this index lists a nondirector, that person's name is not included in the Film Index section. The index to actors and actresses working in the field is useful, although the writer's claim that only those who have made a significant contribution to the genres are included is somewhat inaccurate in view of the listing of people such as Elsie Randolph (a delightful singing and dancing partner to Jack Buchanan in the Thirties, but hardly an actress associated with horror or fantasy films). A similar listing to the Film Index is provided for television productions, and is, perhaps, more useful in view of the difficulty in documenting this type of information. Disappointingly, not even directors are always included here. I know such information is not always easy to find, but I cannot believe there is not some source for documenting, for example, the directors of *Dr. Who.*

The book concludes with a useful Alternate Title Index. Aside from some of the faults already noted, the major problem appears to be that the author does not clearly define what are science fiction, horror, or fantasy films, and, thus, many productions, which are really mysteries, such as *Alias Bulldog Drummond* or *Frenzy*, are included erroneously.

My Hollywood: When Both of Us Were Young by Patsy Ruth Miller/ *The Hunchback of Notre Dame* edited by Philip J. Riley. $49.95. (O'Raghailligh Ltd. Hardcover. 233 pages/208 pages. Index. Illus.)

This is two books for the price of one, being the autobiography of silent screen actress Patsy Ruth Miller and a facsimile reprint of Wallace Worsley's copy of the script for the Lon Chaney vehicle, *The Hunchback of Notre Dame,* in which Miller was featured as Esmeralda. An undeniable amount of love, hard work, and money has gone into the publication of this volume, and any criticism would be unfair. However, let me state that I wish there had been editorial footnoting of the Miller text. The editor should have explained there is no such place as "The Congressional Archive, which supposedly lists all the movies ever made," and that Miller is presumably referring to the Library of Congress, which does not do the latter. Similarly, after Miller notes the death of her last husband, the editor should have indicated the date, or when she mentions a Brussels archive which has the last surviving print of *The Girl I Loved*, the editor should have confirmed that this is, one supposes, the Cinémathèque Royale de Belgique, and that, again one supposes, that archive has also preserved the film. These are minor points, but they would have helped make the volume more useful to the academic or scholarly reader. Patsy Ruth Miller is certainly an entertaining writer, who gives the film buff what he wants to read, but I do wish she had not made that comment comparing pre-Hitler Berlin to contemporary New York City. The filmography by Jeffrey Carrier is a detailed one, and I only spotted one error. What Carrier states is a 1915 Theda Bara version of *The Hunchback* is the 1917 feature, *The Darling of Paris.*

I have no criticisms with regard to the *Hunchback* half of the volume. The documentation here is extraordinary. A first-rate introduction by George Turner is backed up by pages and pages and pages of contemporary material on the film. An excellent editing job by Riley. Recommended.

Radio and Television: A Selected Annotated Bibliography, Supplement Two: 1982-1986 compiled by Peter K. Pringle and Helen E. Clinton. $27.50. (Scarecrow Press. Hardcover. 249 pages. Indexes.)

A companion volume to two earlier books compiled by William E. McCavitt, the present offering contains annotated listings for almost 1,000 English-language publications. A useful work for the librarian or the academic, but of minimal interest to anyone else.

Outcasts: The Image of Journalists in Contemporary Film by Howard Good. $25. (Scarecrow Press. Hardcover. 195 pages. Index. Illus.)

A slim volume (for its price), *Outcasts* offers a scholarly view of the depiction of the press in recent features, plus an historical overview of the subject. Good takes himself very seriously, as does the media about which he writes. Perhaps he should loosen up, particularly as neither journalists nor Hollywood deserve to be treated respectfully.

The Hollywood Hissables by Gregory William Mank. $49.50. (Scarecrow Press. Hardcover. 530 pages. Index. Illus.)

This volume boasts one of the most wretched titles of any film book in recent memory. It consists of *Films in Review* style career articles on nine "villainous" players: Lionel Atwill, John Carradine, Lon Chaney, Jr., Laird Cregar, Brian Donlevy, Claude Rains, Basil Rathbone, George Sanders, and Gale Sondergaard. Unfortunately, the filmographies which accompany each chapter do not contain the amount of documentation one has come to expect from *Films in Review.* The author lacks a certain sensitivity and seems unable to fully discuss the basic tragic qualities of the careers of Cregar and Sanders. Also, either he or the publisher should have updated the manuscript, and recognized that DeWitt Bodeen and Gale Sondergaard are both (sadly) departed from this earth.

May 1989

The Cinema History of Burt Lancaster by David Fury. $29.95. (Artist's Press. Hardcover. 301 pages. Illus.)

A book produced as a labor of love can all too often be nothing more than that: a book which its author/publisher adores, but which fails to excite its readers, let alone a reviewer. Happily, such is not the case with David Fury's *The Cinema History of Burt Lancaster.* Certainly, the author's enthusiasm for Lancaster is such that he cannot find fault with his subject, but, on the whole, it can probably be argued that there is little in Lancaster's career with which to find fault. Looking through a list of the actor's films, it is amazing how many good performances he has turned in through the years. Fury has produced a "Films of" book which is better by far than most other "Films of" books produced to date. He provides a first-class summation of Lancaster's life and work and follows it with a detailed film-by-film overview. The chapters on each film—from *The Killers* in 1946 through *Shoeless Joe* [*Field of Dreams*] in 1989—include complete cast and technical credits, a lengthy plot synopsis, the author's comments on the production, and a selection of contemporary critical commentary from such publications as *Life, The New York Times, Newsweek, Time,* and *Variety* (as well as some which seldom get quoted, including *Commonweal* and *The New Republic*).

Aside from its contents, *The Cinema History of Burt Lancaster* is everything which a good book should be; pages are sewn not glued, and the binding and paper meet the highest of library standards. Recommended.

Flights of Fancy: The Great Fantasy Films by Kenneth Von Gunden. $25.95. (McFarland. Hardcover. 295 pages. Index. Illus.)

The author dismisses other books on the fantasy film as depending for their contents "solely upon the whims of the author," and then proceeds to provide the reader with just such another book. *Flights of Fancy* is nothing more than a collection of lengthy essays on 15 fantasy films, which the author considers of particular importance. There is no new research apparent here, with the author relying upon what he has read in other books and periodical articles on the subject. There are quite a few mistakes, such as calling the 1935 *She* a lost film. If one of the 15 feature films in this book is one of your favorites then perhaps *Flights of Fancy* is for you. Otherwise, don't waste your money.

The Where Book of Movies by Dale Thomajan. $6.95. (Quinlan Press. Paperback. 151 pages. Illus.)

This is a quiz book with all the questions relating to where certain incidents took place; for example "Where does Cary Grant hide John Qualen in *His Girl Friday?*" (In a roll-top desk) or "Where might an American get the blues on a honeymoon holiday in spring?" (Paris). I can't really decide why anyone should want a book such as this, but I suppose it might while away an empty hour or two. Unfortunately, I don't have any empty hours, nor, for that matter, can I answer most of the questions.

The Loud Silents: Origins of the Social Problem Film by Kay Sloan. $24.95. (University of Illinois Press. Hardcover. 166 pages. Index. Illus.)

The Loud Silents is an important volume documenting early silent films from the 'teens and earlier which dealt with such social issues as women's rights, labor, and politics. Ms. Sloan has done her homework, basing her text on primary sources such as trade papers and contemporary volumes. Obviously, this book does not represent all there is to write on the subject, but it is a first-class introduction, well documented for the academic user and well written for the film enthusiast.

Mae West: A Bio-Bibliography by Carol M. Ward. $39.95. (Greenwood Press. Hardcover. 241 pages. Index. Illus.)

Carol M. Ward has done a first-rate job in providing documentation on Mae West's life and career. Her book—the best that I have yet read in Greenwood's Popular Culture Bio-Bibliographies series—contains chapters on Mae West's life and art, reprints of interviews with the actress, notably from *Movie Classic;* a bibliographical essay; a superb bibliography; a chronological biography; and detailed credits for Mae West's film and theatre appearances. Ms. Ward expertly gathers together her material on West, and presents it not only in a scholarly fashion, but also a readable one—and how few academic volumes are actually readable? Recommended.

June 1989

Bette, Rita, and the Rest of My Life by Gary Merrill. $18.95. (Lance Tapley. Hardcover. 273 pages. Illus.)

In this frank and open autobiography, Gary Merrill talks about Bette Davis, to whom he was married, and Rita Hayworth, with whom he had an affair following his breakup with Davis, but he also tells us a great deal about subjects with which most readers will be less familiar. There are amusing anecdotes, well peppered with four-letter words about Merrill's show business career, including *This Is the Army* ("It was quite a sight to see uniformed men strolling hand in hand in the lovely Long Island sunsets") and "those horseshit movies of the Sixties." (Quite frankly, a glance at the list of Merrill's films published at the book's close indicates how few major features he appeared in between 1943 and 1977.) More interestingly, one learns of Merrill's love for the state of Maine. We read of Carl Sandburg, who taught the author that life ought to be spent doing what felt right, not what one thought he ought to do, and who once told a woman he met at a party, "I'm very sorry, my dear, but you bore me, and I don't have enough time left." Above all, one comes to admire Merrill's sincerity. You may not agree with his politics, but he was one of the few men in show business willing to take a stand without consideration of the consequences, be it in Vietnam or Selma, Alabama. And I loved his story of cornering William F. Buckley in an airport telephone booth and yelling at the terrified columnist, "Come on out, you fascist . . . come on out and defend your politics."

If you have read Davis's and B. D. Hyman's stories, you will certainly want to read Merrill's side. If you have not read Davis and Hyman, don't. Let Gary Merrill tell you what really happened. My only regret is that the publisher did not use the five blank pages at the back of the book for an index.

James M. Cain by Paul Skenazy. $18.95. (Crossroad/Ungar/Continuum. Hardcover. 203 pages. Illus. Index.)

Although concerned primarily with James M. Cain, the novelist, Paul Skenazy's book does provide documentation on the classic Film Noir screen adaptations from Cain's works: *Double Indemnity, Mildred Pierce,* and *The Postman Always Rings Twice.* The book notes that the first film to be based on a Cain story was *She Made Her Bed* in 1934, and that Cain received screenplay credit for only three features: *Algiers, Stand Up and Fight,* and *Gypsy Wildcat.* I particularly enjoyed some of Cain's quoted comments on the film industry, such as his description of a film writer as "a prostitute of the arts." Skenazy provides a useful chronology and a complete listing of films based directly or indirectly on Cain's works.

Carl Th. Dreyer edited by Jytte Jensen. $9.95. (Museum of Modern Art. Paperback. 95 pages. Illus.)

To coincide with the first complete retrospective of the films of Danish director Carl Th. Dreyer, the Museum of Modern Art (MOMA) has published a useful and attractive book, the best sections of which are the introduction by Ib Monty and the detailed filmography. The remainder of the volume is taken up with three esoteric essays, a translation of Dreyer's screenplay for *Medea,* and a useful bibliography. In perusing the last, I am reminded that this is not the first pamphlet on the subject published by MOMA; back in 1964 Eileen Bowser authored *The Films of Carl Dreyer.*

Television Studies edited by Gary Burns and Robert J. Thompson. $39.95. (Praeger. Hardcover. 272 pages. Index.)

Unfortunately television has become so popular that it encourages college professors to get together and produce unreadable texts, such as the essays which constitute this book, supposedly remedying "the lack of a forum for current research on television." This is not research—this is self-indulgence. Had the author of the essay on *Dallas* been allowed near the show it would not have lasted one month let alone 11 years. The author of an essay on a show titled *The Young Ones* feels it necessary to explain that one's arse is one's rectum, but does not bother to explain what is the show about which he is writing. Based on the quotes from some of the characters, it sounds like fun, but only by scouring the footnotes did I discover that *The Young Ones* is "a BBC production currently broadcast in the United States on MTV on Sunday nights." (I couldn't find this show listed in my cable TV guide—so much for this book being current.) I have news for the editors of *Television Studies:* there is a good source for current television research. It is called *The Perfect Vision,* published quarterly from Box 357, Sea Cliff, N.Y. 11579, and each issue, which contains far more text than this current offering, costs a mere $6.00. Not Recommended.

Lon Chaney: Master Craftsman of Make-Believe by N. L. Ross. $19.95. (Quality R. J. Press. Hardcover. 288 pages. Illus. Index.)

This book represents a good example of why one should not judge a book (or anything else for that matter) by its cover. Initially, I felt there was something amateurish about this volume. I avoided dealing with it. Well, I should be ashamed of myself. This is a first-rate, well-written, carefully researched biography of Lon Chaney. I might have welcomed additional technical credits in the filmography and I wasn't entirely happy at the way Ross has combined his bibliography with his footnotes, but those quibbles aside I have nothing but praise for this book. We need a good book on Chaney, and this volume more than suffices. Recommended.

The Holt Foreign Film Guide by Ronald Bergan and Robyn Karney. $39.95. (Henry Holt. Hardcover. 638 pages.)

This is a reference guide to more than 2,000 foreign-language films, with the authors providing original and American and/or British release titles, credits, synopses and short comments on each feature. I had a number of reservations with regard to the book. I was sorry that two of my favorite foreign films—*The Last Chance* and *To Forget Venice*—were missing. Many of the synopses are too brief; for example, there is far more to the plot of *Betty Blue* than this volume might suggest. There are a few errors; for example, the English language version of *L'Atlantide* did not utilize the cast of the French version. There are differences of opinion; for example, *Black and White in Color* is a lot better than the authors suggest. There should have been an index, at least to the directors. And there are curious omissions, such as Sarah Bernhardt's *Queen Elizabeth,* so important to film history, or Raymond Bernard's *Miracle of the Wolves.* Having got that off my chest, let me state that this *is* a useful reference book, and it *is* one that I am going to keep near at hand on my reference shelves. Recommended with reservations.

July 1989

Chaplin and American Culture: The Evolution of a Star Image by Charles J. Maland. $29.95. (Princeton University Press. Hardcover. 442 pages. Index. Illus.)

There are many, many books on Chaplin (and this is not the first to appear in this the comedian's centennial year), but *Chaplin and American Culture* takes a new look at an old subject. It studies Chaplin and his relationship with the United States, which it should not be forgotten treated him more than shoddily in refusing to readmit the comedian to this country in 1953. Aside from the scholarship and grace of Maland's writing, this book has the advantage of presenting documentation from the government's files on Chaplin. (Interestingly, Maland reveals that the Immigration and Naturalization Service told the FBI that there was no solid evidence to exclude Chaplin from this country.) *Chaplin and American Culture* concludes with a lengthy chapter on Chaplin's "restoration," in part evidenced by IBM's use of the Little Tramp image as a corporate symbol. I wonder if Chaplin, the Leftist-sympathizer, would have approved of this use of his image? I am sure that Chaplin, the capitalist and millionaire, would have. Recommended.

Eva Le Gallienne: A Bio-Bibliography by Robert A. Schanke. $37.95. (Greenwood Press. Hardcover. 218 pages. Index.)

Eva Le Gallienne's fame is something I find difficult to comprehend. I know many individuals who saw her on the stage in the Twenties or Thirties. I know some who worked with her. All talk glowingly of Miss Le Gallienne as one of the greatest actresses of the century. I saw her on stage only once, in a 1976 touring production of *The Royal Family*, and what I saw and heard was an elderly lady with no voice who simply did not project. It is, perhaps, difficult for me to comprehend why anyone would want to compile or publish a reference book on Eva Le Gallienne, but, having written that, let me state immediately that Robert A. Schanke's volume is a first-rate piece of solid, careful research. This is a basic reference book. A 17-page biography is followed by a detailed listing of Le Gallienne's work in stage, film, radio, and television; an annotated listing of works by Le Gallienne; and a similar listing of books and articles on the actress. Curiously, the entries on Le Gallienne's stage and television appearances contain more information than is given on her screen work. Auteurists and members of the Writers Guild will love this book—Schanke lists the screenplay credit first prior to naming the producer and director.

Film Directors on Directing by John Andrew Gallagher. $45. (Greenwood Press. Hardcover. 313 pages. Index. Illus. A paperback edition is also available at $14.95 from Praeger.)

John Gallagher has compiled a very worthwhile collection of interviews with the following directors: John G. Avildsen, Tony Bill, Michael Cimino, Abel Ferrar, James Glickenhaus, Menahem Golan, Stuart Gordon, Ulu Grosbard, Anthony Harvey, Ted Kotcheff, Adrian Lyne, John Milius, Alan Parker, Franc Roddam, Mark Rydell, Susan Seidelman, Joan Micklin Silver, James Toback, François Truffaut, and Wim Wenders. With only four of these directors the subject of book-length studies, Gallagher's work is automatically of importance. The compilation is noteworthy for the intelligence of the author's questions—he does not sink to such infantile wondering as to what so-and-so was like to work with, or ask questions to which the answers are obvious to any film buff or film student. He is concerned with the making of specific films, and he obtains some hard facts from his interviewees (although he is not above prefacing some queries with his own personal thoughts).

Readers who may complain that this is nothing more than a questions and answers volume should note that Gallagher provides informative introductions to each subject, as well as a filmography, which provides minimal credit information but does offer the unusual information as to the home video source for each title. Recommended.

Show Us Life: Toward a History and Aesthetics of the Committed Documentary edited by Thomas Waugh. $19.50. (Scarecrow Press. Paperback. 536 pages. Index. Illus.)

Originally published in hardcover format in 1984, *Show Us Life* gathers together 24 essays (despite the publisher's claim that the majority have never before been published, all but eight have appeared elsewhere) on the "committed" documentary. This is a worthwhile collection of items, but it raises a number of questions. The first of which is why the term "committed" can only be applied to Leftwing and Communist filmmakers. Are not Rightwing filmmakers just as committed, and should not space be afforded here to such prominent documentary filmmakers as Leni Riefenstahl and to Humphrey Jennings (the latter surely committed as few other filmmakers to a love of his country)? Why do so many of the writers here ignore what is happening around the filmmakers about whom they write? There is a good essay on Esther Shub, yet the writer does not mention that her entire career in the Soviet Union took place during the reign of Josef Stalin. How did she survive and prosper working for a man whose crimes against humanity are equal to those of Adolph Hitler? Presumably, she served Stalin as much as Leni Riefenstahl served Hitler, so why not give both women equal space? Why are so many obscure filmmakers discussed, and probably the best know of Leftist documentary filmmakers, Marcel Ophuls ignored? Perhaps because the editor and writers of this book ignore the reality of life is the reason why the majority of the committed filmmakers about whom they write make films which are seen only by those sympathetic to the cause. If a writer cannot break through the barrier and interest a noncommitted audience in what he has to say, then neither can a documentary filmmaker.

August 1989

Tough Guy: The American Movie Macho by James L. Neibaur. $24.95. (McFarland. Hardcover. 224 pages. Index. Illus.)

Following some brief introductory comments on "The First Movie Tough Guys" in the genres of the Gangster, Western, and Military Films, the author offers career evaluations and filmographies on Edward G. Robinson, James Cagney, Humphrey Bogart, George Raft, Clark Gable, John Wayne, John Garfield, Marlon Brando, James Dean, Elvis Presley, Clint Eastwood, and Sylvester Stallone. (I would never have considered Elvis Presley as an example of "American Movie Macho.") There is nothing new in this book, and it is difficult to understand to whom it would appeal. If readers are particularly interested in any of the subjects, there are book-length studies on all of these individuals readily available.

Cowboy Shooting Stars by John A. Rutherford and Richard B. Smith, III. $7.50. (Empire Publishing. Paperback. 100 pages. Illus.)

This is a useful little publication, providing checklists on all the stars of (as the authors describe them) "Budget" Westerns, from Bob Allen to Whip Wilson. Listings include only film titles, years of release, releasing studios, and running times., Even to someone such as myself, who has little interest in the subject, it seems a handy, little publication to have around. Incidentally, the photograph in the center pages of the Hollywood Sign shows that landmark as it was rebuilt in 1978. It is, therefore, not the Hollywood Sign under which the B-Westerns flourished.

Toms, Coons, Mulattoes, Mammies, & Bucks by Donald Bogle. $15.95. (Continuum. Paperback. 322 pages. Index. Illus.)

This is a new and expanded edition of Bogle's 1973 historical overview of Blacks in American films. It is a readable and useful study, although not as important as Thomas Cripps' *Slow Fade to Black.* It is also concerned primarily with Blacks in mainstream American cinema and devotes only 15 pages to the Black films for Black filmgoers industry. My main criticism is that Bogle does not consider it necessary to provide either a bibliography or any footnotes as to his sources.

Lash LaRue: The King of the Bullwhip by Chuck Thornton and David Rothel. (Empire Publishing. Paperback. 160 pages. Illus.)

I have to admit a total lack of interest in Lash LaRue and his career, although I am intrigued to learn from this book that both Johnny Cash and Waylon Jennings wear black clothing out of childhood admiration for this hero of the "B" Western. I suppose it is a matter of having been reared on the films of cowboy stars such as LaRue, and it makes me ponder as to whether this interest will die out in a few years as the children who grew up on LaRue grow old and die.

My lack of interest notwithstanding, I am happy to recommend *Lash LaRue: The King of the Bullwhip*. It is an entertaining volume, full of fascinating documentation, including an interview with the subject, an interview with Phil Smoot, who directed LaRue in two features from the Eighties—*Dark Power* and *Alien Outlaw*—and a strangely incomplete filmography. (Why doesn't it include the two aforementioned features?) Fans of Lash LaRue will love this book. Others will find it a fascinating glimpse into the world of the "B" Western fan, a world which we cannot possibly hope to understand or be a part of.

Them Ornery Mitchum Boys by John Mitchum. $11.95. (Creatures at Large Press. Paperback. 378 pages. Index. Illus.)

Robert Mitchum's younger brother, John, has written a really splendid, anecdotal history—it is far more than an autobiography—of the two men's lives and careers. One senses that this is a book written by a real man about real people and real happenings. Even if you don't know, let alone care, about some of the people that Mitchum is discussing, you are still fascinated and engrossed by his comments. I loved Mitchum's stories of Monty Woolley, Orry-Kelly, Anatole Litvak and Paulette Goddard, and I wish I could repeat them here. Robert Mitchum comes across both as a fun-loving jerk and a sensitive individual. He once greeted a visitor in the nude, bent over and remarked, "Look at the Movie Queen!" When a crew from *Entertainment Tonight* approached him to comment on John Huston's death, he said bitterly, "I should have told them if they were so hot to make his death a part of their 'entertainment,' they should drive an oak stake through his heart. It would've been the only way to make sure that great soul was gone." This is a book which I am looking forward to rereading at my leisure. Highly Recommended.

The Fox That Got Away by Stephen M. Silverman. $19.95. (Lyle Stuart. Hardcover. 356 pages. Index. Illus.)

A fascinating account of the last days of the Zanuck dynasty at 20th Century-Fox, when Darryl was head of the studio and his son, Richard, head of production, *The Fox That Got Away* takes the reader into the political world of Hollywood filmmaking. If nothing else, it shows how power corrupts and the foolishness of ever putting any of your thoughts into memos which writers such as Silverman might one day read. There's some juicy gossip here, particularly concerning Bella Darvi's sexual relationship to both Darryl and Virginia Zanuck, and a similar relationship between the two Zanucks and an unnamed but easily identifiable "pretty boy Fox contract player in the early 1950s." The book is not particularly well-written and Silverman appears to have nothing more than a passing knowledge of film history.

Horror: A Connoisseur's Guide to Literature and Film by Leonard Wolf. $27.95. (Facts on File. Hardcover. 262 pages. Index. Illus.)

Because this is a guide for connoisseurs, as opposed to a regular, old guide for the likes of you and me, the film listings are selective. They have been chosen, the author claims, for "historical spread, thematic variety and fun." The latter two claims I can accept. The first is questionable, for a book which does not include the earliest film version of *Frankenstein* and has Lon Chaney represented by only one feature, *The Phantom of the Opera* (coincidentally, the only version of *The Phantom* to be included). Each listing runs several paragraphs, and includes credit information, although the author does not know the difference between production companies and distributors. Thus,

20th Century-Fox, the distributor, is listed as the production company for *Invaders from Mars,* while the producer, Edward L. Alperson, is not mentioned at all. There are factual errors; for example, the song, "Raindrops Keep Falling on My Head" is noted by the author as being featured in *A Clockwork Orange,* while the song is actually "Singin' in the Rain." Leonard Wolf can and does come up with a nice turn of phrase from time to time, but he is also remarkably sloppy in his use of words, as in the entry for the 1931 *Frankenstein,* where the last two paragraphs begin "Finally. . . . "

I cannot honestly accept this book as a reference guide. It is a highly personal, highly selective commentary on books and films of the horror genre and has literary rather than reference merit. Incidentally, the author claims to be a native of Transylvania who now lives in New York City (where else?). Based on the photograph which accompanies his biography, I would not like to meet him on a dark night.

A Moving Picture Feast: The Filmgoer's Hemingway edited by Charles M. Oliver. $37.95. (Praeger. Hardcover. 186 pages. Index.)

This is the third book to consider screen adaptations of Ernest Hemingway's works, and it is a disappointing volume. Comprised of papers read at a conference at Boston's John F. Kennedy Library in May 1981, *A Moving Picture Feast* contributes perhaps a little to Hemingway scholarship but almost nothing to film scholarship. The fifteen writers responsible for the essays here have almost no understanding of the film industry or of film history. One writer, Frank M. Laurence, labors under the misapprehension that the director is responsible for deciding how a film should be released. One wishes that the editor, who appears to have had little to do, could have at least contributed a detailed filmography (there is none), or a bibliography which lists more film reviews than those from the obvious sources, such as *The New York Times. A Moving Picture Feast* left me hungry.

September 1989

Hibernian Green on the Silver Screen by Joseph M. Curran. $39.95. (Greenwood Press. Hardcover. 172 pages. Index. Illus.)

Last year, McFarland published a book by me, titled *The Cinema and Ireland,* which dealt with film production in that country, along with the Irish image on American screens and the Irish in Hollywood. I was, therefore, most interested in reading Joseph M. Curran's work, *Hibernian Green on the Silver Screen,* which is described by its publisher as "a pioneering work" (which it is not), and which considers the last two areas previously written about by me in *The Cinema and Ireland.* Curran mentions neither my book, nor two other volumes on the motion picture and Ireland which have since appeared. Whether Curran's failure to acknowledge these other texts is deliberate or based on a lack of knowledge on the subject I do not know. Either way, it emphasizes the shortcomings of his book. On the positive side, Curran does provide useful data on Irish history from the last century onwards. On the negative side, he offers virtually no new research or documentation on the subject, and he displays only a passing knowledge of film history. He has limited his research to only two major film-related libraries and examined the papers of only one filmmaker relevant to the subject, John Ford. He has made no attempt to interview anyone associated with his area of study and utilized only texts which are readily available.

The section on the silent era is highly selective. The author barely mentions the work of the Kalem Company in Ireland and is ignorant of the fact that Sidney Olcott visited Ireland each year between 1910 and 1914, not just in those two years. An important production such as *Ireland a Nation* is treated in a cavalier fashion, and Curran does not even name its producer. The chapter on John Ford fails to consider the director's pose as a professional Irishman and the negative response which his films have evoked in Ireland. To suggest that in *The Quiet Man,* Ford makes "stage Irishmen" seem real is ludicrous. The third episode from *The Rising of the Moon* has no similarity to Lady Gregory's play of that name, and the author should have stated such. "A Minute's Wait" episode, also from that production, is an insult to the Irish people. Has Curran ever seen this film? If so, why not attack it as a "striking example of negative Irish stereotyping," instead of leveling such criticism at *Ragtime?* Throughout, one has the impression that the author is writing of films which he has not seen and about which his opinions are derived from secondary sources (and often not always the best of sources).

I could continue with such criticism, but I would only be accused of bias in favor of my own book. Such is the lot of a book reviewer!

Movie Musicals on Record compiled by Richard Chigley-Lynch. $39.95. (Greenwood Press. Hardcover. 392 pages. Index.)

Subtitled "A Directory of Recordings of Motion Picture Musicals, 1927-1987," this book is a companion volume to the compiler's 1987 *Broadway on Record.* Entries are in alphabetical order by film title and list the record(s) with catalog number, composers, cast, and songs included on the record. There is also a chronological listing of the musicals, plus indexes by performer and composer/lyricist. I felt certain this book could not be complete, but I could find no omissions, although I am dubious that all 78 rpm recordings are included. *Movie Musicals on Record* is a book which is a "must" for serious collectors of recordings and recommended also for anyone interested in the musical film.

Double Vision: My Life in Film by Andrzej Wajda. $19.95. (Henry Holt. Hardcover. 136 pages. Index. Illus.)

The author is one of the better known Polish directors, whose features, including *Generation, Kanal, Ashes and Diamonds,* and *Lotna* were once considered exciting, new classics of the foreign film, but have received little attention in recent years. This slim volume is neither an autobiography nor a treatise on filmmaking, but rather the director's views on the production of films as it has affected his life since the late Forties. Wajda's style is very easy to read; it is not esoteric and, indeed, is almost simplistic at times. He is proud of his country's films and of the Solidarity movement. He despairs that "There is no longer a cinema that addresses itself to everyone." Was there ever? Wajda reveals quite a bit about himself and his colleagues, particularly actor Zbysek Cybulski who was considered the James Dean of Polish cinema until his death in 1967. About the filmmaking process, he reveals less, discussing it as if his readers are totally unfamiliar with even the basics of production.

Tom Mix Highlights by Andy Woytowich. (Empire Publishing. Paperback. 52 pages. Illus.)

This is basically a collection of sketches of Tom Mix, all of which have a nice period feel to them, both in terms of the graphics and the accompanying copy. There is Tom Mix with his parents, in London, promoting the Straight Shooters Club, and so on. Aside from the sketches, the book contains a title list of Mix's films, a one-paragraph foreword by Paul E. Mix, and a one-page biographical commentary by M. G. Norris. The last commences with the description of Mix as "The greatest cowboy film star of all time." This is a book for all those willing to take that description at face value.

The Official Splatter Movie Guide by John McCarty. $10.95. (St. Martin's Press. Paperback. 145 pages. Illus.)

Oh dear! Do I really want to review a book by a man who is described on the back cover as "the John Simon, the Leonard Maltin, the Siskel and Ebert of carnography?" If I am honest with myself, yes, I do, because I rather enjoyed leafing through these pages documenting *The Female Butcher* or *Surf Nazis Must Die.* Over 400 "splatter" features are included here, with credits and commentary on each. Some, such as *Jaws,* I am surprised to find listed in the category. Splatter Movies as a genre appear to date back to 1968 and *Night of the Living Dead.* I wonder if any other film genre has regressed in quality as has this one—or is it that a deterioration in the quality of the "splatter" movies is an improvement in the eyes of the genre's enthusiasts? Recommended.

Movie Memories edited by Sharida Rizzuto. $18.60 for 4 issues. (Baker Street Publications. Paperback. 110 pages. Illus.)

I have been sent a review copy of the second issue of this mimeographed journal, which contains generally slight articles on personalities from Hollywood's golden years, some book reviews, fan club listings, and poetry. Some names familiar to *Classic Images* readers, such as Michael Ankerich and L.E. Ward, are here. The writing is basically nonprofessional and must be accepted as such. The poetry is extraordinary in its quality. This is not fan magazine poetry—this is the real thing. One of the poems, "Homage to Garbo" by Carl Brennan, is a superb work. It was so surprising to find it and the other pieces here. What a pity this publisher cannot issue a journal devoted entirely to poetry such as this. That same publisher, Baker Street, notes that its magazines are published irregularly. Perhaps it should change its name to the Baker Street Irregulars?

Television Network Prime-Time Programming, 1948-1988 by Mitchell E. Shapiro. $45. (McFarland. Hardcover. 743 pages. Index. Illus.)

This is an excellent supplementary text to Tim Brooks and Earle Marsh's *The Complete Directory to Prime Time Network TV Shows 1946-Present* (first published in 1979)—the information in the two books does not overlap. Shapiro provides day-by-day charts and listings not only for ABC, CBS, and NBC, but also DuMont and the Fox network. (Actually, Fox only makes it to the Saturday and Sunday schedules.) There are also daily programming summaries notating series premieres and key programming moves. Thus, the reader may learn that on Sunday in 1948, ABC introduced such choice shows as *Stained Glass Windows* and *Sing-Co-Pation,* while in 1988, the same network had progressed to *Buck James, Dolly,* and *Supercarrier.* And the critics wonder why American television cannot produce a series like Dennis Potter's *The Singing Detective!* Shapiro's *Television Network Prime-Time Programming, 1948-1988* is an obvious "must" for all library collections and well worth purchasing by all wealthy TV buffs who can afford the price. Recommended.

No Name on the Bullet: A Biography of Audie Murphy by Don Graham. $19.95. (Viking Press. Hardcover. 396 pages. Index. Illus.)

Audie Murphy single-handedly killed 240 Germans—215 more than Sergeant York massacred during the First World War. If Murphy had not killed those men, any one of them might have killed him.

Audie Murphy received the most medals ever awarded an American soldier, including the Congressional Medal of Honor, and yet he felt that all the soldiers who died during the war were far more deserving of those medals. He had nightmares about the violence in which he had participated, but once he looked at a group of Universal executives eating in the commissary and remarked, "With one hand grenade, a person could get rid of all these no-talent bastards at one stroke." He had a baby face, but practiced cruel and often violent practical jokes on his fellow workers, and always carried a gun. Audie Murphy disclaimed any ability to act, but directors, critics, and public alike found much to admire in his films.

In other words, Audie Murphy was a mass of contradictions. He is hard to understand, and, as the years have gone by since his 1971 death, his work and his place as an American hero has become harder and harder to accept. This new biography makes it very plain that its author regards Murphy as a great man, a great Texan, and a genuine war hero, unlike a recent American president who fought the Second World War in the backlots of Burbank and Culver City. There is evidence of tremendous research into all aspects of Murphy's life and career, but, unfortunately, the biography falls short in that the author's style lacks fire and the vigorous emotional strength necessary to tell its hero's story. With that reservation, *No Name on the Bullet* is still well worth reading, particularly by those, such as myself, who may be inclined to dismiss Audie Murphy as someone who has no place in today's society or today's thinking.

October 1989

Trigger Remembered by William Witney. $12.95. (Earl Blair Enterprises. Paperback. 78 pages. Index. Illus.)

Quite a few directors have written books, but I am sure that William Witney is the first director to have written a volume about a horse. But what a horse! Witney first met Trigger, and his lesser half Roy Rogers, in 1938, after the former had made his screen debut as Olivia De Havilland's mount in *The Adventures of Robin Hood.* In a fascinating and beautifully produced book—the color photograph of

Trigger and Roy on the back cover is just gorgeous—Witney documents Trigger's career through the end, with a bonus chapter on Roy Roger's dog, Bullet. Witney makes no personal commentary regarding Trigger's being stuffed and displayed at the Roy Rogers Museum in Victorville. Personally, I find it an inglorious end to a glorious career, although I take solace in knowing that although his body may be on display to the tourists, Trigger's soul has found freedom in eternity. I should also add here that Witney also tells us much about his own work and colleagues, which makes his book doubly recommended.

Film Noir by Bruce Crowther. $14.95. (Frederick Ungar/Continuum. Paperback. 192 pages. Index. Illus.)

This present work does not in any way supplant the two most important books on the genre, *The Dark City: The Film Noir* by Spencer Shelby and *Film Noir* by Alain Silver and Elizabeth Ward but it does provide a useful, and highly readable, introduction to the subject. Happily, the author does not consider film noir a strictly Forties and early Fifties phenomenon, and writes of recent films as being examples of the genre. His approach to the films is not always from an auteur's viewpoint. Thus, he will devote much space to the stars rather than the director or screenwriter, and, for example, he makes three mentions of the 1948 classic *Pitfall*, but fails to note that its director is Andre de Toth. *Film Noir* was originally published in the United Kingdom, and the text uses English phrases and English spellings.

Radical Innocence: A Critical Study of the Hollywood Ten by Bernard F. Dick. $27. (The University Press of Kentucky. Hardcover. 264 pages. Index. Illus.)

Billy Wilder once joked of the Hollywood Ten, also known as the Unfriendly Ten, "only two had any talent; the other eight were just unfriendly." I am inclined to go along with Mr. Wilder. The author of *Radical Innocence* is not, and he claims that an examination of the ten's writings, screenplays, and films, divorced from their politics, proves him correct. Certainly, Mr. Dick presents considerable research to support his position. His chapters on Samuel Ornitz, Lester Cole,

John Howard Lawson, Herbert Biberman, Albert Maltz, Alvah Bessie, Adrian Scott, Edward Dmytryk, Ring Lardner, Jr., and Dalton Trumbo are quite obviously well researched and documented, and his book also provides extremely useful biographical data, including listings of published works and films. Unfortunately, Mr. Dick is sometimes vague as to who are the producers of such films, crediting several simply as independent productions, and does not seem to be aware that United Artists is a distributor not a producer. He is also inclined to credit too much to the writer and not enough to the director. Thus, Andre de Toth—this is my month for promoting Mr. de Toth—is not mentioned in connection with *None Shall Escape,* while screenwriter Lester Cole receives full credit. Yet it is obvious that de Toth's experience as a newsreel cameraman at the start of the war is the basis for at least one sequence in the feature. Recommended with reservations.

Syndicated Television: The First Forty Years, 1947-1987 by Hal Erickson. $45. (McFarland. Hardcover. 418 pages. Index. Illus.)

It is amazing how many different configurations authors and publishers can come up with in an effort to produce new reference books on television. I wish they would all get together and compile one, definitive volume documenting all television programs from the pioneering years to the present.

The latest, as its title indicates, deals with syndicated television programs, which are covered only randomly in Tim Brooks and Earle Marsh's *The Complete Directory to Prime Time Network TV Shows, 1946-Present.* The current volume documents syndicated series by decade and by subject: Adventure/Mystery; Children's; Comedy; Drama; Game/Quiz; Informational; Music/Variety; Religious; Specials, Miniseries and Mini-Networks; Sports; Talk/Interview; Westerns; and Women's. The information provided here is generally hard to find elsewhere, and certainly not in books, and one appreciates the compiler's efforts. For example, I don't know where else one could look for an entry on the Lutheran Church's *This is the Life* or *Lee Marvin's Lawbreakers.* The book also includes a first-rate introductory essay explaining syndication and a useful annotated bibliography. My only criticism/question is why the book's title suggests that the entries start in 1947, whereas there is no entry for any show earlier than 1950. Highly Recommended.

The Ruby Slippers of Oz by Rhys Thomas. $14.95. (Tale Weaver Publishing. Paperback. 232 pages. Illus.)

Ostensibly, this is the story of what happened to Judy Garland's ruby slippers from *The Wizard of Oz*—and a fascinating story it is. But *The Ruby Slippers of Oz* is also a moving and often sad account of a group of gay men in Hollywood, all of whom were besotted with the thought of owning those slippers. The author offers what is I believe the first examination of movie memorabilia collectors and what makes them behave as they do. Certainly, it is not always a pretty saga, but it's a highly readable and engrossing one.

Footage 89 edited by Richard Prelinger and Celeste R. Hoffnar. $89. (Prelinger Associates, Inc. Paperback. 796 pages. Index.)

Footage 89 is a regrettably and often irritatingly flawed attempt to provide documentation on all film collections in the United States, Canada, Mexico, the British Virgin Islands, and Cuba. Entries are arranged by state, and include addresses, contact people, descriptions of collections, and information as to cataloging, access, licensing, viewing, and duplication facilities, etc. A series of articles at the front of the volume deal with various aspects of the commercial film library industry, including history, copyright, research, and preservation. A list of researchers is appended with addresses and credits. A subject index at the back of the volume might have been useful had the subject headings been more akin to those used by the library community, and had the headings listed all the libraries with footage on a particular subject. As it is, only a chance reading of the introduction to the subject index indicates that some fifty or so libraries have holdings on most of the subjects and so they have not been listed under individual headings.

I don't want to minimize the work of the book's compilers, who have undertaken a tremendous task. Nothing like this has been contemplated or accomplished before, and the book will have a lasting value. The problem is the lack of editorial control. The editors seem content to include any information provided to them and are unable to differentiate between the importance of one collection compared to another. Thus, a major stock footage library with several million feet of film on file, such as the Los Angeles-based Cameo, is given the same amount of space as a New York collection with one film which is not even under its direct control. Some libraries list the subjects they hold; others do not. At least one company is permitted to include the advertising slogans applicable to the feature films in its collection. Producers are listed simply because they are producers, and there is no indication under a producer's name that its stock footage is handled by another library. A number of major studio stock footage libraries—Lorimar, CBS/MTM and Warner Bros.—are not included. Worst of all, and I think an embarrassment to the editor, is that he uses the book to overpromote his own company and that of an associate, Petrified Films. Thus, the latter and Prelinger Associates get a combined coverage of 24 pages, while Sherman Grinberg Film Libraries, the largest stock footage house in the United States, garners a mere 3 pages.

It is, I believe, the editor's intention that this book should appear on an annual basis. Hopefully, the next edition will take into account many of these criticisms. If it does, I shall have no hesitation in recommending it without reservations. As it is, I can only acknowledge its worth, and wish for improvement.

Cinema 2: The Time-Image by Gilles Deleuze. $14.95/$39.50. (University of Minnesota Press. Paperback/Hardcover. 344 pages. Index.)

According to its back cover, *Cinema 2* "brings to completion Gilles Deleuze's work on the theoretical implications of the cinematographic image." This is a book for those who cannot communicate in plain English and for academics who do not have the remotest idea of how a film is made, what a director does, or the functions of an optical printer. In other words, it is not for you or me. If you live in Minnesota, please obtain a copy and send it to your state senator as an example of how Minnesota wastes public money. If you live outside Minnesota, mourn for the tree which gave its life that this trash could come into being. Oh, I nearly forgot. The book is translated by Hugh Tomlinson and Robert Galeta. Into what language, the publisher does not indicate.

Looney Tunes and Merrie Melodies by Jerry Beck and Will Friedwald. $14.95. (Henry Holt. Paperback. 390 pages. Index. Illus.)

This is a totally revised and apparently rewritten edition of a book published a few years ago by the Scarecrow Press, although the present volume gives no indication as to its antecedent. It is a unique reference source, providing year-by-year documentation on all the Warner Bros. cartoons from 1930 through 1988, together with informative listings on Warner Bros. cartoon TV specials and feature films, the Bugs Bunny TV show, and a final "That's Not All" roundup documenting other Warner Bros. cartoon contributions (largely to the war effort).

Each entry begins with the name(s) of the cartoon's central character, the animators, production supervisors, music arrangers, etc. (the credits increase as the cartoons get more recent). The narrative includes a plot synopsis, historical information, and reference to the music used. Recommended.

Color Collectors Guide edited by Robert Brosch. $7.95. (Archival Photography. Paperback. 81 pages. Illus.)

This is number one of what is intended to be an annual publication. It consists primarily of color reproductions of hundreds of lobby cards from horror and science fiction films, copies of which are available for sale from the publisher. It also includes a foreword by Forrest J. Ackerman, a highly opinionated piece on collecting lobby cards by the editor, an article on lobby cards from 3-D films by Jerry Ehardt, and innumerable black-and-white photographs. It is literally brimming over with material, so much so that it took me a couple of browsing sessions to figure out what it was all about. Unfortunately the section headed "Please Read This Section First" explains the abbreviations used to identify the various types of motion picture advertising matter, but does not tell the reader what to expect. Nevertheless, this is a fascinating publication, and certainly a "must" for the movie memorabilia collector. It has nothing to do with the subject to hand, but I particularly enjoyed the collection of photographs from *Dead of Night*, and the editor's comments thereon.

Beyond Ballyhoo: Motion Picture Promotion and Gimmicks by Mark Thomas McGee. $25.95. (McFarland. Hardcover. 237 pages. Index. Illus.)

I have often thought there is a need for a book on the promotional efforts used to "sell" films to the public. The present volume is not quite what I had in mind, but it's a good start. McGee's style is somewhat flippant and he glosses over the early years of film promotion—indeed the silent era is barely mentioned and it is amazing that he can ignore the spectacular premier of *Hell's Angels. Beyond Ballyhoo* is a good introduction to the subject, but we still need a definitive academic study on the topic.

November 1989

Blind in One Ear by Patrick Macnee and Marie Cameron. $19.95. (Mercury House. Hardcover. 298 pages. Index. Illus.)

Patrick Macnee is best known for his work in the television series, *The Avengers,* and when I received his autobiography my first reaction was what had he done about which anyone would be interested in reading. The answer, even Mr. Macnee admits, is not a great deal as far as a career is concerned, but his autobiography is entertaining and considerably outrageous, particularly when he writes of his mother and her lady friend, "Uncle" Evelyn who took away young Macnee's trousers and made him wear a kilt. Macnee spares the reader nothing in writing of his life, his loves, and his failures. There are times when one can sympathize with him and times when one wishes he had kept it all to himself. Had I not had to read *Blind in One Ear,* I certainly would not have done so. Now that I have, I am pleased that duty prevailed.

Teenagers and Teenpics: The Juvenilization of American Movies in the 1950s by Thomas Doherty. $12.95. (Unwin Hyman. Paperback. 275 pages. Index. Illus.)

Any volume which can compare Edwin S. Porter to Sam Katzman—"Be the first, not the best"—deserves to be noticed. And Thomas Doherty's book is a superb, academic study of the American cinema's initial pandering to adolescent taste in the Fifties—a trait which has, regrettably, continued into the Eighties. This study is well considered, well documented and well written, taking into account both earlier historical developments and sociological movements in society. The book also includes a useful filmography of Fifties teenpics, identified by content category, such as "Clean Teenpic" or "Weirdie." Highly Recommended.

Hot Toddy by Andy Edmonds. $19.95. (William Morrow. Hardcover. 242 pages. Index. Illus.)

This purports to be the true story of how actress Thelma Todd died, with the author maintaining it was a particularly vicious end as a result of her ties to organized crime rather than accidentally at the hands of director Roland West as has generally been assumed. Any book of this type should include footnotes or chapter-by-chapter documentation supporting the published statements, but Ms. Edmonds provides none. She is able to provide lengthy quoted conversations, despite neither she nor a tape recorder being present at the time. The author makes a number of factual errors when documenting Todd's film career in the Twenties and without any compunction whatsoever makes unproven assertions regarding various film figures: Thomas Ince's death was an unsolved murder, Robert Harron's death was linked to his homosexuality, Valentino was either bisexual or homosexual, etc.

Hot Toddy is a book which most film buffs are going to want to read. I wish it could have been a better book, because, quite obviously, the truth of Thelma Todd's death needs to be told, and also because here was an actress who deserves a detailed career study.

Rating the Movies. $13.95. (Publications International. Paperback. 512 pages. Illus.)

Prepared by the editors of *Consumer Guide* (not the better known *Consumer Reports*), *Rating the Movies* has entries on more than 3,700 feature films, with each entry consisting of a one-paragraph write-up, minimal credits, MPAA rating, Consumer Guide rating, and information as to Academy Award nominations or wins. Both American and foreign films are included, but there is no indication as to how the chosen films were selected, nor are any writers credited, nor any explanation offered as to how the ratings were arrived at. The entries are bland and easy to read, and if you are not very demanding in your reading matter, this is for you.

Your Guide To Having Recorded Media Appraised by David C. Miller. $20. (CCM Associates. Paperback. 31 pages.)

This is a very disappointing and very overpriced guide to what is involved in having a collection of films, movie memorabilia, etc., appraised. With the IRS tightening the regulations with regard to tax writeoffs on donated collections, a guide such as this could be valuable. Unfortunately, this present offering is little more than a promotion for the appraisal company of CMM Associates and does not, for example, even consider the major question of setting a fair market value for collections of copyrighted films which cannot legally be sold.

Harlow by Irving Shulman. $12.95. (Mercury House. Paperback. 408 pages. Illus.)

This is one of the volumes in a new—and excellent—series of reprints, "The Lively Arts Series," published by Mercury House. When *Harlow* first appeared back in 1964, it was criticized as a muckraking biography of one of the screen's most glamorous personalities. Since then, a good many real muckraking biographies have been published, and I think it is about time to recognize Irving Shulman's book for what it is, a first-rate, well-researched biography, written with considerable flair and style. I only wish the new publishers had been willing to prepare an index, something which this book sadly lacks. Recommended.

Grand National, Producers Releasing Corporation, and Screen Guild/Lippert: Complete Filmographies with Studio Histories by Ted Okuda. $35. (McFarland. Hardcover. 241 pages. Index. Illus.)

Ted Okuda is one of the better of the younger historians documenting lesser aspects of the film industry's past. He has already prepared checklists of the films of Monogram (in 1987) and the Columbia shorts (in 1986). His new book provides complete filmographies, together with brief corporate histories, for Grand National, Producers Releasing Corporation, and Screen Guild/Lippert. Each entry includes producer, director, screenplay, and players credits, along with running time, release date and a one-paragraph synopsis. The documentation found in this volume is hard to locate elsewhere, and the book is a valuable addition to any library. Recommended.

The Bride of Frankenstein edited by Philip J. Riley. $19.95. (Magic-Image Filmbooks. Paperback. Illus.)

This volume, described as "Universal Filmscripts Series, Classic Horror Films—Volume 2," contains the original shooting script for *The Bride of Frankenstein,* and much, much more. It is a very worthwhile project, marred by a number of poor judgments in terms of overall layout. A lacklustre and rather unnecessary history of Universal is followed by a foreword and introduction, both of which, by virtue of their titles, should obviously have come at the book's beginning. The foreword by Forrest J. Ackerman is quite useful and interesting and should, perhaps, have been merged with Gregory Wm. Mank's informative production notes, which appear later. The introduction by Valerie Hobson is not an introduction at all, but an interview, which again is quite informative despite some rather uninspiring questions. It might have been advisable to cut the questions out and edit together Miss Hobson's comments as a narrative. The script, reproduced in its entirety, is claimed to be James Whale's personal copy, and its value would have been enhanced had the script been notated as to differences (if any) between it and the film as released. Additional supplementary material makes the book extremely valuable, and despite my reservations, *The Bride of Frankenstein* is recommended.

The Boys: The Cinematic World of Laurel and Hardy by Scott Allen Nollen. $22.95. (McFarland. Hardcover. 153 pages. Index. Illus.)

The Boys—nice title!—is yet another book on Laurel and Hardy. Despite more than worthwhile earlier volumes by Charles Barr, William K. Everson, John McCabe, and Randy Skretveldt, this new book does offer some originality. The author analyzes the work of Laurel and Hardy in an easily readable fashion and discusses their films as they apply to the comedians as a couple, and in relationship to physical objects, animals, women, and society. A filmography, which takes up almost one third of the book, is somewhat unnecessary in view of what has been published before, but it does rate each film and also describes the character relationship between the two men in each production.

Cinema in Iran, 1900-1979 by M. Ali Issari. $42.50. (Scarecrow Press. Hardcover. 446 pages. Index. Illus.)

With Iran in the news with depressing regularity, this, the first book-length study in English of that nation's film industry is particularly welcome. The book covers Iranian cinema up to the coming to power of the Ayatollah Khomeini and conveniently glosses over what Khomeini's return to Iran must have done to its film industry. Similarly, an extremely useful thirty-page historical and cultural introduction to Iran does not concern itself with what the Ayatollah has wrought. *Cinema in Iran*'s text is supported by assorted tables, statistics, and maps. Recommended.

Mask and Scene: An Introduction to a World View of Theatre by Diana Devlin. $25. (Scarecrow Press. Hardcover. 227 pages. Index. Illus.)

This book by the granddaughter of British actor Lewis Casson is so simply constructed that I am puzzled as to whether it is intended for teenager readers rather than adults. Whatever its intended audience, it is a useful introduction to world theatre for anyone desiring nothing more than a passing knowledge of the subject.

December 1989

Good Night, Sweet Prince: The Life and Times of John Barrymore by Gene Fowler. $12.95. (Mercury House. Paperback. 477 pages. Index. Illus.)

Another in the Lively Arts series, *Good Night, Sweet Prince* was first published in 1944, some two years after the death of its subject, John Barrymore. Gene Fowler ingratiates himself into the text, perhaps with reason because he was one of the actor's close friends—indeed, that friendship helped provide Fowler with access to Barrymore's personal papers. I suspect many will find this book hard to read. It is written in a style not usually associated with today's biographical efforts, and despite the grace and flow of Fowler's phrases, it seems rather creaky. Despite such reservations, it is good to see this first major biography of a great actor of the stage and screen back in print.

Eleanor Parker: Woman of a Thousand Faces by Doug McClelland. $32.50. (Scarecrow Press. Hardcover. 291 pages. Index. Illus.)

Doug McClelland has compiled a remarkable amount of documentation on actress Eleanor Parker. Included here is a biographical essay, professional credits for Miss Parker's work in all media, information on unrealized film, television, and stage projects, and every imaginable type of bibliographic data. The 47-page listing of published references to the actress is an extraordinary labor of scholarship and love. Eleanor Parker is lucky to have a writer/historian such as McClelland to document her career, while the latter is equally fortunate in having an actress such as Miss Parker on whom to lavish such attention. Recommended.

Confessions of an Actor by Laurence Olivier. $8.95. (Penguin Books. Paperback. 361 pages. Index. Illus.)

First published in 1982, Sir Laurence Olivier's autobiography is now reprinted in what passes these days for a cheap paperback. Quite obviously, Olivier devotes more pages to his work in the theatre than elsewhere, explaining, "the wide media of films and television do not usually tax one's energies beyond their normal capacities," whereas "the playing of a great stage role does exactly this." However, he is candid and extremely detailed in his commentary on all aspects of a long and honored career. The book closes with a select chronology, beginning with the actor's birth, but, curiously, the publisher of this reprint did not consider it necessary to close that section with Olivier's 1989 death.

Max von Sydow: From the Seventh Seal to Pelle the Conqueror by Peter Cowie. $10. (The Swedish Film Institute. Paperback. 87 pages. Illus.)

Published by the Swedish film journal, *Chaplin,* in association with the Swedish Film Institute, this monograph appears simultaneously in Swedish and English. It is a first-rate work, lucidly and informatively written, with von Sydow's own comments on his career interspersed through the text. A basic filmography concludes the book. (The volume may be ordered directly for $10, including postage, but readers should send an international money order, not a personal check.) Recommended.

From the Inkwells of Hollywood compiled by Charles and Pat Searle. 3 volumes. $42. (Searle's Autographs. Paperback. Unpaged. Illus.)

These three mimeographed catalogs are primarily intended to offer for sale autographs of stage, screen, radio, and television personalities. However, they are important in that they provide immediate information as to the values of such autographs, as well as facsimiles, for authentication purposes, of the autographs in question. Volume 1 covers A-L, Volume 2 M-Z, and Volume 3 Academy Award winners.

Israeli Cinema: East/West and the Politics of Representation by Ella Shohat. $30/$13.95. (University of Texas Press. Hardcover/Paperback. 312 pages. Index. Illus.)

Aside from Ora Gloria Jacon-Arzooni's *The Israeli Cinema* (published in 1983), this is the first history of Israeli cinema to appear in the United States, and it is a major achievement. The book provides an overview of filmmaking in Israel, from the pioneering years of Palestinian cinema at the beginning of this century through the present. Every fictional film produced in Israel is discussed, and my one regret is that the author did not see fit to include some form of filmography.

Ella Shohat is sometimes overly wordy, but most of the text is easily readable and informative. Every effort is made to place Israeli cinema in its relationship to the politics of the country. One learns a considerable amount about the diversity of Israeli and Jewish culture, as well as the various genres into which the relatively small number of Israeli features can be divided. Recommended.

The Bear by Daniele Heymann. $9.95. (St. Martin's Press. Hardcover. 52 pages. Illus.)

This is the book of the film by Jean-Jacques Arnaud, which is scheduled to open in the U. S. after this review is written. If, as the publisher hopes, the film becomes "one of the most-talked about, most beloved movies of 1989," then there will presumably be a market for this storybook. If not, it is doubtful it will have much appeal for film buffs or for children who will find it a little too sophisticated. The color photographs of the bear are nice.

Black Action Films by James Robert Parish and George H. Hill. $35. (McFarland. Hardcover. 408 pages. Index. Illus.)

This volume provides casts, credits, synopses, and critical commentaries on more than 225 Black action or "blaxploitation" features and made-for-television movies produced over the last four decades. The "Rocky" films are also included, although I would not have considered them part of the genre. In his foreword, Richard Roundtree claims that a book such as this is long overdue. Perhaps. But I find it hard to get enthusiastic over reading about so much that is little more than mediocre and often represents some of the worst in American cinema.

TV Movies and Video Guide edited by Leonard Maltin. $5.95. (New American Library. Paperback. 1,277 pages.)

Now in its 20th year of publication, *TV Movies and Video Guide* has become one of the most useful and best known of all film reference books. Along with Ephraim Katz's *The Film Encyclopedia,* it is the book that everybody involved in film, either professionally or as a hobby, must have on his or her desk. Incidentally, at $5.95, it is also the best bargain in film books around.

The (1990) edition contains entries on 18,500 features and made-for-TV movies. Each entry provides information on year of release, running time, director and leading players, along with a one-sentence comment which tells the reader not only what the film is about but also what Maltin thinks of it. There is also a Maltin rating for every feature, with which the reader may not always agree—I certainly think *Broken Blossoms* is not rated high enough—but at least here one knows that the rating has merit in view of the respectability and reputation of the rator. One nitpicking complaint is that when I first flipped through the book I wondered what the symbol beside some of the entries meant. The "Key to This Book" did not tell me (as it should have), and only by reading the introduction did I discover it meant the film was available on video (another useful feature of *TV Movies and Video Guide*). Highly Recommended.

The Films of Hopalong Cassidy by Francis M. Nevins. $19.95. (The World of Yesterday. Paperback. 324 pages. Illus.)

Francis Nevins is one of the leading figures in the world of popular culture and the author of a recent biography of Cornell Woolrich. I mention these facts because the author's reputation and stature is such that he has produced a book far superior than its title or its publisher might suggest. Nevins is a serious writer, as evidenced by the following comment on Hopalong Cassidy creator Clarence Mulford: "It was Darwin, and the social Darwinian thinkers like Herbert Spencer, and the documents of nineteenth-century paganism like the Fitzgerald translation of Omar Khayyam, that shaped Mulford's vision of the Western hero." That is not the type of commentary one usually finds in a volume devoted to a Western hero. Not only is the text superior, but also the book boasts a detailed filmography. My only complaints are that there is no index and some of the captions to the illustrations suggest a lack of knowledge of the plots of the films. Also, based on the number of his photographs, either the publisher or the author has a fixation for Jimmy Ellison. Recommended.

Orson Welles: A Biography by Barbara Leaming. $10.95. (Penguin Books. Paperback. 562 pages. Index. Illus.)

This is the paperback edition of a book first published in 1985. Two biographies of Orson Welles appeared that year, one by Leaming and one by Charles Higham. The Leaming biography perhaps has more style, but the Higham book is probably closer to the reality of Welles' life and career. Leaming received Welles' cooperation, and seems to have believed most of what he told her without reservations.

The Magic Lantern by Ingmar Bergman. $10.95. (Penguin Books. Paperback. 308 pages. Index. Illus.)

Regardless of what one may think of Ingmar Bergman as a director, his autobiography makes fascinating reading in that the man comes across as totally arrogant, repulsive, and antisocial. Bergman's attitudes towards his parents, his women, and his children are deeply disturbing—perhaps the reason why his films can often be distasteful.

Charlie Chan at the Movies by Ken Hanke. $29.95. (McFarland. Hardcover. 288 pages. Index. Illus.)

Here is more than one would ever want to know about Charlie Chan. Hanke divides his book into six sections: Warner Oland as Chan; Sidney Toler as Chan at 20th Century-Fox; Sidney Toler as Chan at Monogram; Roland Winters as Chan; Imitations and Offshoots (Mr. Moto and Mr. Wong); and Later Chans. Each of the films is discussed in detail, with complete credits, and an introduction explains how author Earl Derr Biggers first brought Charlie Chan to life in *The House without a Key.* (Unfortunately, the author makes a mistake on the second page of his book in claiming, incorrectly, that *Seven Keys to Baldpate* was first filmed in 1926.)

Cinema Sequels and Remakes, 1903-1987 by Robert A. Nowlan and Gwendolyn Wright Nowlan. $75. (McFarland. Hardcover. 966 pages. Index. Illus.)

At least two earlier volumes have attempted to document film sequels and remakes, but the current work is certainly the most detailed (despite the authors' admittance that many films have been omitted). Unfortunately, the present book is neither the most accurate nor the most reliable in terms of its authors' opinions, which are obviously not based on a great knowledge of film history nor viewing of many of the features in question. Basic credits are provided for both the original film and its sequels and remakes, except for those films which date from the 'teens, for which, generally (and without explanation) no credit information is supplied. To admirers of Clive Brook's *On Approval* (of which I am one), it will come as a surprise to discover that this is "an atypical farce about 'silly ass' upperclass Englishmen." Equally surprising is the authors' statement that Miriam Hopkins ends *Becky Sharp* facing "a life of growing old alone with her beauty gone." Those sort of half-assed comments and mistakes typify too much of the book for it to qualify as a serious work of reference, despite the obvious immense amount of work that must have gone into its compilation.

1990

January 1990

Julie Andrews: A Bio-Bibliography by Les Spindle. $37.95. (Greenwood Press. Hardcover. 163 pages. Index. Illus.)

Fans of Julie Andrews will doubtless welcome this book, documenting her career on stage, screen, radio, and television, and what comments I make as to its academic failings will have little impact. There is a lot of valuable information here, interspersed through a very dull text, but there are also a lot of problems with the book. For example, I can well recall as a child listening to Andrews on the popular radio show, *Educating Archie,* starring ventriloquist Peter Brough and his dummy, Archie Andrews (the British equivalent of Edgar Bergen and Charlie McCarthy). Despite the show's gaining national recognition for Julie Andrews, it is mentioned only in passing, and Peter Brough is not named at all. The only reviews quoted in the filmography are from *The New York Times* and *Variety,* reviews from both of which are well indexed and easily accessible in anthologies at most large public libraries. No page numbers are listed for any articles in the bibliography. Several articles are listed as merely cover stories, with no indication as to titles or authors. Six reviews are listed for *That's Life,* although without the names of their authors, while none are listed for *Mary Poppins.* Finally, as the author is not presumably on intimate terms with Miss Andrews, I wish he had not chosen to refer gushingly to her throughout the text as "Julie."

George Raft by Lewis Yablonsky. $12.95. (Mercury House. Paperback. 288 pages. Index. Illus.)

Dean Martin told the author of this biography of a party at George Raft's home in which the women swam nude and, at dinner, found one hundred dollar bills tucked in their napkins. "George had class," says Martin. If this is your idea of class, or if you can sympathize with a sobbing George Raft, clutching his grandmother's rosary, while awaiting sentencing for tax evasion, this book is for you. The author makes no pretense of being unbiased in his study—he is a fan, and despite a wealth of detail and some fascinating stories, this is obviously a book which should not be accepted at face value. First published in 1974, *George Raft* is reprinted as a quality paperback in the Lively Arts series, but neither the publisher nor the author (if he is still alive) has made any effort to update the book in any way, and so the actor's story ends in the early Seventies with him very much alive and endorsing songs about "fags" and saying "expletive-deleted you" to the Queen of England.

New German Cinema: A History by Thomas Elsaesser. $50/$14.95. (Rutgers University Press. Hardcover/Paperback. 430 pages. Index. Illus.)

New German Cinema is an excellent, detailed, and well-considered history of West German filmmaking from the end of the Second World War to the present. Despite the title, East German filmmaking is not considered, and the author devotes little space to the Forties and Fifties, only really getting down to analyzing the growth of German filmmaking from the Seventies onwards. He discusses German cinema as both an art form and an industry, paying particular attention to government subsidies. Complementing the volume is a filmography and a directory of new German filmmakers. Highly Recommended.

Encyclopedia of American War Films by Larry Langman and Ed Borg. $95. (Garland. Hardcover. 696 pages. Illus.)

This is a weighty and expensive tome, providing basic credit information together with a synopsis/commentary on almost 2,000 war films, along with additional general essays on topics such as the cold war and concentration camps. I am sure that not all war films have been included, and I am disturbed that a two-page list of sources used by the authors indicates that only secondary ones have been utilized and that many contemporary items have been ignored unless they happen to have been reprinted in anthologies. I also do not understand how a book such as this can have been compiled without use of Craig Campbell's indispensable *Reel America and World War I* (published by McFarland in 1985). One useful item in the book is an appendix of wars and their related films. There are wars here of which I have never heard!

Some additional thoughts of a bemused book reviewer: According to the title page, this is Vol. 873 in the Garland Reference Library of the Humanities, which makes me wonder what humanity has to do with war? This book is printed on "acid-free, 250-year-life paper," guaranteed not to survive the next war. This is the first book which Garland has agreed to send this department for review—it will probably be the last.

The First Film Makers by Richard Dyer MacCann. $22.50/$12.50. (Scarecrow Press in association with Image & Idea. Hardcover/Paperback. 313 pages. Index. Illus.)

Certainly not new viewpoints on the lives and work of Griffith, Ince, and von Stroheim, as the compiler suggests, because all of the essays here are reprinted from other sources—some of them quite old—but a useful companion with which to while away a few hours, and equally meritorious as a textbook (although I doubt there are many film courses concentrating on such worthwhile a topic as the early years of American cinema).

In view of the compiler's lavish and rather overdone praise of critic Richard Schickel's work, it is not surprising that the publisher uses equally lavish praise from Schickel with which to "sell" the book.

Horrors! by Drake Douglas. $22.50. (The Overlook Press. Hardcover. 418 pages. Illus.)

This is a reprint of a book first published in 1966, and now available with a new introduction and epilogue. *Horrors!* is a book which is hard to classify. As its author maintains, it is neither a history nor a reference work, but "simply a book to be read and enjoyed." On that basis, it meets the author's criteria, with chapters on the werewolf, the monster, the mummy, the walking dead, the schizophrenic, the phantom, and the hunchback, plus two chapters on "the creators of horror." Unfortunately, there is something slightly pretentious about the book and its author (who prefers to write under a pseudonym). It is also a little difficult to appreciate how anyone can compile a volume on the horror film and admit, without apology, that they have never seen Lon Chaney in *The Phantom of the Opera.*

If you want a guide to horror films, there are many more worthwhile volumes available. If you just want a book to pick up and read for a while at your leisure, *Horrors!* will serve your purpose.

3-D Movies by R. M. Hayes. $45. (McFarland. Hardcover. 414 pages. Index. Illus.)

How great to come upon a book which fills a real need, and fills it well. *3-D Movies* provides an excellent history of the genre from 1903 to the present and also includes a survey of current 3-D processes and services, as well as a filmography and critical commentary (providing *full* credits and intelligent observations). My only complaint is the lack of a bibliography. Highly Recommended.

The Cinema of Stanley Kubrick by Norman Kagan. $12.95. (Continuum. Paperback. 249 pages. Illus.)

First published in 1972, and expanded here to include new chapters on *Barry Lyndon, The Shining,* and *Full Metal Jacket, The Cinema of Stanley Kubrick* is a first-rate survey of the director's work. Following a couple of general chapters, a chapter is devoted to each feature, with a detailed summary and analysis, and discussion of recurring motifs. A filmography completes the volume, but why no bibliography and no index?

Glancing through the list of Frederick Ungar/Continuum books, I am reminded what a fine group of film publications this publisher is offering.

Motion Picture Copyrights & Renewals, 1950-1959 by David Pierce. $89. (Milestone. Paperback. 494 pages.)

Film Superlist: Motion Pictures in the U. S. Public Domain, 1950-1959 by Walter E. Hurst and William Storm Hale. $500. (Seven Arts Press. Hardcover. Unpaged.)

The basic concept for each of these books is the same. The authors have taken the 1950-1959 catalog of motion picture copyright registrations, published by the Library of Congress, and annotated each entry with information as to which film or television program has been renewed, listing the renewal number. To assist users, Pierce has marked each renewal with a black dot, while Hurst and Hale have written "N" beside all those not renewed. Pierce has added an abbreviation for the name of the individual or organization renewing the copyright and appended a key to the abbreviations at the front of his book. Hurst and Hale have taken the index to registrants and indicated here also which copyrights have been renewed.

Pierce has added a 24-page supplement to his book, listing films from the decade which were registered after 1959, and the copyrights in which were also renewed. Those distributors who sell *The Red Balloon* as a public domain title will be embarrassed to note its presence in this list! Hurst and Hale provide considerable, additional materials, including copies of various Copyright Office circulars, a vast amount of documentation explaining the reliability or otherwise of the Copyright Office, and a fascinating sampling of copyright search reports.

Both books are highly recommended. Obviously, Pierce's volume is more affordable and also looks somewhat cleaner and neater than the Hurst/Hale volume. I don't think any self-respecting film-oriented library or individual involved in the marketing of public domain titles can afford to be without either volume. (The Hurst/Hale volume, of course, follows the same format and binding as the two earlier Film Superlists published by Seven Arts Press.)

Razzle Dazzle: The Life and Work of Bob Fosse by Kevin Boyd Grubb. $29.95. (St. Martin's Press. Hardcover. 292 pages. Index. Illus.)

If you want to accept the concept of the auteur, there are only two figures in recent film history to whom that word can be applied. They are Woody Allen and Bob Fosse, the subject of this first-rate autobiography. With the possible exception of *Star 80,* all of Fosse's films are brilliant, if sometimes flawed works: *Sweet Charity, Cabaret, Lenny,* and, of course, *All That Jazz.* His contributions to the New York musical stage are equally memorable—*Sweet Charity, Pippin, Dancin',* etc.—all belonging to an era in American musical theatre which appears to have passed. As the author notes, "Today, people don't have to dance, the scenery does."

Razzle Dazzle examines all of Bob Fosse's works, with commentaries by many of his associates, and, without ever giving the appearance of muckraking, it carefully documents his personal life and the many women with whom he was sexually involved. The title *Razzle Dazzle* perhaps gives the initial impression of a superficial work, but the book might equally be subtitled "Nitty Gritty" in that its author has obviously undertaken tremendous research in his efforts to bring the life and career of a brilliant dancer, choreographer, and director into its true perspective. Recommended.

The Children's Hour: Radio Programs for Children, 1929-1956 by Marilyn Lawrence Boemer. $25. (Scarecrow Press. Hardcover. 230 pages.)

A charming collection of essays on 47 radio programs for children, which aired between 1929 and 1956, with the author attempting to recapture the pleasure which listeners must have felt upon the initial broadcast.

A House with Four Rooms by Rumer Godden. $18.95. (William Morrow. Hardcover. 319 pages. Illus.)

The second volume of novelist Godden's autobiography includes interesting comments on screen adaptations of her work, notably Jean Renoir's *The River* and Michael Powell's *Black Narcissus.* Of the latter, she writes, "There is not an atom of truth in the film."

February 1990

Iris: A Journal of Theory on Image and Sound. $8 per issue. (Institute for Cinema and Culture. Paperback Journal published biannually.)

Published in France since 1982, *Iris* now has one of its two annual issues published each year in the USA effective with No. 9. That particular issue is concerned with "Cinema and Cognitive Psychology," which is apparently an audacious American response to Lacanian psychoanalysis. Enough said!

Film Review, 1989-90 edited by F. Maurice Speed and James Cameron-Wilson. (W. H. Allen. Paperback/Hardcover. 192 pages. Index. Illus.)

Published in the United Kingdom and available from specialized American bookstores, the 45th edition of this popular annual contains all the usual departments: releases of the year, in memoriam, film books of the year, the ten most promising faces of the year, video releases, and "Letter from Hollywood" by Anthony Slide. Enough said!

Hollywood Greats of the Golden Years: The Late Stars of the 1920s through the 1950s by J. G. Ellrod. $27.50. (McFarland. Hardcover. 240 pages. Index. Illus.)

A couple of poorly written introductory paragraphs and a listing of film titles for the more obvious and popular of deceased stars is all that this feeble excuse for a book has to offer. Enough said!

Children's Live-Action Musical Films: A Critical Survey and Filmography by Thomas J. Harris. $24.95. (McFarland. Hardcover. 208 pages. Index. Illus.)

The author provides detailed credits together with essays on the following features: *The Wizard of Oz, Hans Christian Andersen, The*

5000 Fingers of Dr. T, tom thumb, Babes in Toyland, Mary Poppins, Doctor Doolittle, Chitty Chitty Bang Bang, Pufnstuf, Willy Wonka and the Chocolate Factory, Bedknobs and Broomsticks, Alice's Adventures in Wonderland, The Little Prince, The Slipper and the Rose: The Story of Cinderella, and *Pete's Dragon.* All of the essays are highly readable and entertaining, although I do not always agree with Harris's opinions—I really like *The Slipper and the Rose,* for example—and a number of comments by individuals associated with the productions are included. A videography completes the book.

United Artists: The Company That Changed the Film Industry by Tino Balio. $32.50. (University of Wisconsin Press. Hardcover. 446 pages. Index. Illus.)

Tino Balio continues his history of United Artists—the first volume, published in 1976, covered the company from its founding in 1919 through 1951—with a detailed examination of this preeminent distributor from the early Fifties up to and including its 1980 merger with M-G-M. Along the way, Balio discusses Eagle-Lion (founded in 1946), out of which emerged UA's new leaders, Robert S. Benjamin and Arthur B. Krim, and Lopert Pictures (which UA acquired in 1958). *United Artists: The Company That Changed the Film Industry* also provides a listing of all UA domestic releases from 1951-1978 and a listing of the company's principal producers, with their films, for the same period.

Despite some criticism in *Daily Variety* that Balio was short-sighted in his research and ignored the contributions of some lesser individuals, it is obvious that *United Artists: The Company That Changed the Film Industry* is a superior piece of academic research. Some individuals may have been omitted, but the facts are, quite obviously, here, and presented in an intelligent and readable fashion. Highly Recommended.

Remembering Charlie by Jerry Epstein. $29. (Doubleday. Hardcover. 288 pages. Index. Illus.)

Jerry Epstein knew Chaplin from 1946—he is the credited producer on *The Countess from Hong Kong*—until the comedian's death, and he provides some fascinating insights into Chaplin's later years. At times, he is too easy to forgive the comedian his faults and there are many pages where Epstein is serving as an apologist. Nevertheless, *Remembering Charlie* is a good read and replete with new information, not to mention new photographs. Recommended.

The Tom Mix Book by M. G. "Bud" Norris. $24.95. (The World of Yesterday. Paperback. 379 pages. Illus.)

Merle Norris is a Tom Mix fan, so much so that he obtained a part in the 20th Century-Fox feature, *Brubaker,* just so that he could claim to have worked at the same studio as his hero. Merle Norris is also a dedicated researcher, and his book is full of just about every imaginable fact relating to Tom Mix. There is a complete, detailed filmography, a history of Mix's radio programs, chapters on Tom Mix Little Big Books, Tom Mix comics, Tom Mix sheet music, etc., etc. The book also includes a narrative history of Mix's life, plus an interview with Jane and Eva Novak, both of whom were leading ladies to the cowboy star.

There are a few errors or curious observations. It is hard to understand why *Variety* reviews of Mix's Selig films are included, with a note that the periodical only covered a few, when probably all of Mix's early films were reviewed in *The Moving Picture World, Motion Picture News* or *The New York Dramatic Mirror.* Also, I believe that new research indicates that *Ranch Life in the Great Southwest* is not Mix's first film.

However, these criticisms are small in comparison to the amount of praise which should be heaped on a volume such as this. In a straightforward, down-to-earth fashion, Merle Norris has provided us with everything we could want to know about Tom Mix. I congratulate him and recommend *The Tom Mix Book.*

James Mason: Odd Man Out by Sheridan Morley. $18.95. (Harper & Row. Hardcover. 200 pages. Index. Illus.)

As the author feels it necessary to note several times in his text, this is the first biography of James Mason. It is also, like much of Sheridan Morley's work, rich in style but short in substance. A biography of James Mason should be twice this length and much, much more than this catalog of film titles.

Enchanted Drawings: The History of Animation by Charles Solomon. $75. (Alfred A. Knopf. Hardcover. 322 pages. Index. Illus.)

A beautiful book for everyone who cares about the animated film, *Enchanted Drawings* combines a scholarly text (but there should have been footnotes and more documentation) with an exquisite layout of black-and-white and color photographs. Recommended.

The Post-Feminist Hollywood Actress: Biographies and Filmographies of Stars Born after 1939 by Kerry Segrave and Linda Martin. $39.95. (McFarland. Hardcover. 336 pages. Index. Illus.)

This is basically a collection of short essays—very similar to those which appear in *Current Biography*—on some fifty actresses who were born in 1940 or later and whose career successes date from after the 1963 publication of Betty Friedan's *The Feminine Mystique.* Following each essay is a useful bibliography and a filmography. No effort has been made to interview the actresses in question and although the essays are written from a feminist perspective, there is no indication as to how the actresses feel about their categorization as "post-feminist."

One has to admire authors who come up with new ideas for books, and certainly Segrave and Martin are breaking new ground. There is a particularly good introduction to the subject of female players, and here and at the book's conclusion, the authors have included some fascinating statistics on "gender differences in films." Unfortunately, the book is dated—more than a year out-of-date—in that no mention is made of Candice Bergen's popular television series, *Murphy Brown,* Sigourney Weaver's starring role in *Gorillas in the Mist,* of Jamie Lee Curtis' series, *Anything But Love,* or Jodie Foster's being an Academy Award winner.

March 1990

Highlights and Shadows by Charles G. Clarke. $29.50. (Scarecrow Press. Hardcover. 284 pages. Index. Illus.)

No. 21 in the Filmmakers series, *Highlights and Shadows* is the autobiography of cinematographer Charles G. Clarke, whose career ran from 1915 through 1962. It is the story of a "nuts and bolts" cameraman, always reliable, renowned for his location and second unit work, and under contract to 20th Century-Fox from 1938 until his retirement.

The Cinema as Art by Ralph Stephenson and Guy Phelps. $8.95. (Penguin Books. Paperback. 272 pages. Index. Illus.)

This is a revised edition of a book first published in 1965, under the authorship of Stephenson and Ralph Debrix. It is remarkable in that it is not only a readable introduction to the subject of film as an art form, but also moderately entertaining. However, in terms of historical accuracy, it is somewhat unreliable; for example, *The Black Pirate* is described as the only surviving two-color Technicolor feature and Lee De Forest is called Lee Forrest. *The Cinema as Art* can certainly be recommended as a textbook for high school seniors and undergraduates and will undoubtedly educate the general reader to a wider understanding of the potential of the motion picture, without too much effort on his or her part.

A Variable Harvest: Essays and Reviews of Film and Literature by Jon Tuska. $35. (McFarland. Hardcover. 371 pages. Index. Illus.)

Mr. Tuska has gathered together here a collection of seven film-related essays and twelve literary studies published elsewhere over the past fifteen years. The film topics covered are motley: articles on *Trader Horn, Rain,* and *War of the Worlds,* an overview of the American Western Cinema, and pieces on Yakima Canutt, Spencer Gordon Bennet and Dick Richards. Mr. Tuska lacks neither egotism nor pretention. He is also very well read and lets his audience know it.

Despite such faults (if faults they be), the essays are exceptional in their content and vision and certainly deserving of recapture in a book such as this.

The American West from Fiction (1823-1976) into Film (1909-1986) by Jim Hitt. $29.95. (McFarland. Hardcover. 374 pages. Index. Illus.)

The adaptation of Western novels and short stories for the screen is more than adequately discussed by Jim Hitt, who succinctly notes what elements are lacking from the screen adaptations and whether such a lack is detrimental to the finished work. The book also includes an excellent appendix listing Western authors (with birth and death dates) and film adaptations of their works (with dates of publication, producers, and release years). Recommended.

Hollywood's Unsolved Mysteries by John Austin. $18.95. (Shapolsky. Hardcover. Illus. 277 pages.)

The author commences his book with a 1935 quote which is as relevant today as it was back then: "There is not a reporter in Hollywood who could not rock the country by sitting down at his typewriter and recording merely a portion of the things he knows." And for those who would doubt such a statement they have only to question why Rock Hudson's homosexuality was "covered up" by the entertainment media for almost thirty years. John Austin's thesis is that crimes in the film community are similarly covered up by the Hollywood oligarchy. As he discusses the deaths of Vicki Morgan, Marilyn Monroe, William Holden, Roy Radin, Bob Crane, Natalie Wood, Freddie Prinze, Gig Young, Inger Stevens, and George Reeves, it is hard not to accept Austin's accusations. Unfortunately, the writing style is not as appealing as the thesis, and there is little documentation to back up the author's claims. (Incidentally, is this the first published claim that Reeves was murdered on orders of Eddie Mannix?) Slight, sleazy, and fun reading.

Film at Wit's End by Stan Brakhage. $20. (McPherson and Company. Hardcover. Index. Illus. 183 pages.)

Here, the well-known independent filmmaker Stan Brakhage writes intelligent, warm, and often highly personal tributes to eight of his fellow filmmakers: Jerome Hill, Marie Menken, Sidney Peterson, James Broughton, Maya Deren, Christopher MacLaine, Bruce Conner, and Ken Jacobs. The book's value is increased by an excellent series of filmographies compiled by Robert Haller. Recommended.

Animated TV Specials: The Complete Directory to the First Twenty-Five Years, 1962-1987 by George W. Woolery. $59.50. (Scarecrow Press. Hardcover. Index. Illus.)

As its subtitle indicates, *Animated TV Specials* provides detailed credits and commentaries on some 434 animated television specials—everything is here from *The Care Bears in the Land without Feelings* (1983) to *Uncle Sam Magoo* (1970). Appendices list the most frequently aired shows, those utilizing stop-motion animated puppets, those which were part of a series, plus a breakdown of specials by holidays and subjects. Mr. Woolery is to be congratulated on the amount and quality of his research and scholarship. What a pity the specials themselves did not exhibit this same high quality. Recommended.

Richard Widmark: A Bio-Bibliography by Kim Holston. $35. (Greenwood Press. Hardcover. 160 pages. Index. Illus.)

As with other Bio-Bibliographies in the Performing Arts series, this one provides a chronology; radio, stage and television credits; an annotated filmography; a listing of the actor's films available on video; and an annotated bibliography. Unlike the last bio-bibliography reviewed in this column, Kim Holston's work is almost exemplary. He has provided a first-rate scholarly overview of Richard Widmark's career, and my only complaint is the paucity of the author's personal comments on the films and his failure to provide a more balanced group of reviews rather than relying on just one or two sources. Recommended.

Silent Witnesses: Russian Films 1908-1919, researched and coordinated by Yuri Tsivian; edited by Paolo Cherchi Usai, Lorenzo Codelli, Carlo Montanaro, and David Robinson. (British Film Institute/Edizioni Biblioteca dell'Immagine. Hardcover. 621 pages. Index. Illus.)

A major contribution to film scholarship, *Silent Witnesses* provides a wealth of documentation (including complete credits and detailed synopses) on approximately 300 pre-Soviet Russian films—examples of Czarist cinema—together with lengthy biographies on the major figures of the period. No such task has ever been undertaken before, and it says much for the calibre of its contributors that the book is without fault. No U. S. price is listed; the book retails in Italy for 70,000 lire. Highly Recommended.

I Remember Television: A Memoir by Ira Skutch. $29.50. (Scarecrow Press. Hardcover. 245 pages. Index. Illus.)

This autobiography, No. 8 in the Directors Guild of America oral history series, documents the beginnings of Skutch's working career in the theatre and radio and his many productive years in television from 1945 onward. The book is important to television history, although Skutch is not always the most scintillating of writers. He does offer fascinating word portraits of Fred Coe, Politics in Fifties television (Philco actually showed some courage in standing up to red-baiting), Jackie Gleason, Merv Griffin, Johnny Olson, Nipsey Russell, and, above all, Goodson-Todman Productions (although there is a hint of animosity in Skutch's discussion of Mark Goodson). Skutch's name may not be known to many readers despite his having directed or produced some 10,000 TV episodes. Notwithstanding, this is a book which anyone interested in television history should read and enjoy.

LA 411, edited by Janek Boniecki. $50. (LA 411 Publishing. Paperback. 535 pages. Index.)

LA 411 remains the preeminent directory for everyone involved in film production in Los Angeles. Now published in two volumes, the first book includes listings for advertising agencies, production companies, sets and stages, support services, camera and sound equipment, grip and lighting equipment, props and wardrobe, and post-production. A second volume lists various categories of crew members. The publication is beginning to look a little pretentious and perhaps a return to a more basic format would decrease the number of pages and the price.

April 1990

The Cinema of Adventure, Romance & Terror edited by George E. Turner. $39.95, including postage and handling. (ASC Press. Hardcover. 298 pages. Index. Illus.)

For the last few years, the highlight of each issue of *American Cinematographer* has been a historical article on a specific film. Such articles are models of research and documentation, always accompanied by the most detailed credit information. Some 24 of these articles are gathered together here in book form. The majority are by *American Cinematographer* editor George E. Turner; others are by Rudy Behlmer, Scott McQueen, Paul Mandell, George J. Mitchell, and Michael H. Price. There is neither rhyme nor reason to the selection which includes pieces on *The Hunchback of Notre Dame, All Quiet on the Western Front, White Zombie, Gunga Din, Rope,* and *Night of the Hunter.* It is simply a group of fascinating essays, any of which is deserving of collection in a book such as this. Highly Recommended.

Marilyn: Mon Amour by André de Dienes. $15.95. (St. Martin's Press. Paperback. Unpaged. Illus.)

First published in hardcover in 1986, *Marilyn: Mon Amour* is a beautiful collection of André de Dienes' photographs of the actress taken between 1945 and 1953. The fragile innocence, the zest for life, all the elements which helped make Marilyn Monroe a star, are immediately visible in these photographs, as does the harsher quality of the later photographs warn of the tragedy to come. The text is simple and sincere. The photographs are stunning.

Italian Cinema: From Neorealism to the Present by Peter Bondanella. $17.95. (Continuum. Paperback. 500 pages. Index. Illus.)

The lack of a good, solid English-language history of Italian cinema is surprising—I suppose the best general survey is Pierre Leprohon's *The Italian Cinema* (published in 1972). The best study of post-war Italian cinema is Peter Bondanella's *Italian Cinema: From Neorealism to the Present,* first published in 1983 and reprinted and revised in this new paperback edition. Concentrating on filmmakers and their films, rather than the industry, Bondanella offers an exhaustive study of his subject. His writing style is precise and readable, and he offers useful, brief descriptions of all the major films. His opinions are occasionally dubious; for example, does the Taviani brothers' unintentionally hilarious *Good Morning, Babylon* really deserve three pages of serious consideration? The only major fault lies with an index which is not designed for use by a neophyte. Film titles are not listed in alphabetical order but appear under the names of their directors, implying both that readers should know in advance the name of a film's maker and also that a film has only one. A detailed bibliography and rental information for 16mm prints and videocassettes completes the book. Recommended.

Backstory: Interviews with Screenwriters of Hollywood's Golden Age by Pat McGilligan. $12.95. (University of California Press. Paperback. 382 pages. Index. Illus.)

First published in hardcover in 1986 and now available in paperback, *Backstory* offers lengthy interviews by McGilligan and others with fifteen major screenwriters, many of them now deceased: Charles Bennett, W. R. Burnett, Niven Busch, James M. Cain, Leonore Coffee, Philip Dunne, Julius J. Epstein, Frances Goodrich and Albert Hackett, Norman Krasna, John Lee Mahin, Richard Maibaum, Casey Robinson, Allan Scott, and Donald Ogden Stewart. The comments are often outrageous and always assertive. Bennett denigrates Hitchcock's contributions to his scripts. W. R. Burnett is tough on Communists. Many dispute the script credit claims of their colleagues. It is all quite fascinating and totally compelling reading. However, shame on the author and publisher for claiming the Motion Picture

Country House and Hospital is "a rest home for the penniless and the alone-in-life." It is neither. Recommended.

King Vidor, American by Raymond Durgnat and Scott Simmon. $32.50. (University of California Press. Hardcover. 382 pages. Index. Illus.)

This is a good critical and historical survey of King Vidor's films, marred by the near unreadable sentences of which Raymond Durgnat is so fond. The book also includes a complete filmography.

Alice Faye: A Bio-Bibliography by Barry Rivadue. $35. (Greenwood Press. Hardcover. 239 pages. Index. Illus.)

At the risk of offending Miss Faye's countless fans, I cannot help but wonder if she deserves such a book. I am not trying to denigrate the actress's work, but does it really warrant this type of scholarly documentation? Having written that, I must state that this is a superb work of reference. Included here is a concise biography, a complete and detailed filmography (with a nice assortment of contemporary reviews), a discography, listings of Miss Faye's work on stage and in broadcasting, a list of products which she has endorsed, an annotated bibliography, and much more. Only the list of archival sources warrants criticism for its failure to list the holdings of the Margaret Herrick Library of the Academy of Motion Picture Arts and Sciences and other West Coast sources. I hope that all future volumes in the Bio-Bibliography series will be this good. Recommended.

The Complete Actors' Television Credits, 1948-1988, Volume 1: Actors by James Robert Parish and Vincent Terrace. $59.50. (Scarecrow Press. Hardcover. 560 pages. Illus.)

A second edition of a valuable reference work, the title of the book tells it all. Under each actor's name, the compilers have provided the title of the program (with an indication as to whether it is a series or a pilot), and the date and network on which it aired. 1,587 actors are included, but there are omissions of actors whose credits are perhaps inconsiderable. For example, I looked in vain for William Bakewell and Norman Lloyd. Recommended.

Audiovisual Equipment and Materials: A Basic Repair and Maintenance Manual by Don Schroeder and Gary Lare. $20. (Scarecrow Press. Paperback. 133 pages. Index. Illus.)

I have always believed that if a piece of equipment breaks down you get in an expert to repair it. If this is not your attitude, you will welcome this new edition of a 1979 guide to maintenance and repair of television receivers, video monitors, videocassette recorders, videodisc and compact disc players, and computers.

How to Go to the Movies by Quentin Crisp. $15.95. (St. Martin's Press. Hardcover. 224 pages.)

The general readership will perhaps know Quentin Crisp as the author of *The Naked Civil Servant.* Gay readers will be familiar with his movie column in *Christopher Street.* A collection of Mr. Crisp's pieces from the latter monthly magazine have been gathered together here, and a stylish, if at times flighty, group of essays they prove to be. Mr. Crisp (I am adopting his policy of referring to everyone as either Mr. or Miss) displays a penchant for the witty phrase, such as his description of Joan Crawford, "Age could not wither her nor custom stale her infinite monotony" or of Meryl Streep in *Out of Africa,* "playing the part of yet another extremely annoying woman." All of these pieces transcend simple film criticism and reach a high level of literary quality in a field where such is not particularly a known commodity.

I suppose if one was unkind, one might call these the writings of a rather silly old queen, but then at least they are the honest writings of someone who makes no pretence as to what he is, unlike so many chroniclers of Hollywood history and Hollywood biographies who live in a closet and name others as homosexual in an accusatory fashion.

Karloff and Lugosi: The Story of a Haunting Collaboration by Gregory William Mank. $35. (McFarland. Hardcover. 384 pages. Index. Illus.)

In his latest venture into the field of screen horror, Gregory William Mank writes in considerable and reliable detail on the careers of Boris Karloff and Bela Lugosi, noting that the former enjoyed the greatest success and happiness but that Lugosi memorabilia remains the more collectible. Replete with interviews with a number of co-workers, the book pays particular attention to the eight features in which Karloff and Lugosi appeared together. These films receive the more detailed attention in the filmography, but the work of the two actors on stage, screen, radio, and television is also documented. Mank writes in a straightforward, down-to-earth style, but he does lapse occasionally into fanciful descriptions, such as calling the British industrial city of Dudley "a sooty village." The book boasts a magnificent collection of photographic illustrations, most of which are new to me. I regret the lack of a bibliography, which I know would have included Arthur Lennig's 1974 biography of Bela Lugosi, *The Count,* from which I am sure some of the factual data is taken.

Deadly Resurrection by John McCarty. $3.95. (St. Martin's Press. Paperback. 264 pages.)

The man who coined the phrase "splatter movies," and wrote the 1984 and 1989 books on the subject, has authored his first novel. It is a book which gains in improbability with each new chapter, but still manages to entertain.

The Great Science Fiction Pictures II by James Robert Parish and Michael R. Pitts. $49.50. (Scarecrow Press. Hardcover. 499 pages. Illus.)

This follow-up to the same authors' 1977 volume includes entries on 400 feature films. Fairly detailed credits are provided for each title, along with a commentary which includes a plot synopsis. There is also a basic credit listing for science fiction programs on radio and television. In that the book begins with *Abbott and Costello Meet Dr. Jekyll and Mr. Hyde* and ends with *Zone Troopers,* it is obvious that the authors' definition of "Great" may not necessarily coincide with that of some readers. It is also regrettable that the authors tend too often to quote the comments of other contemporary writers rather than give the reader their own opinions. Despite such reservations, *Great Science Fiction Pictures II* is a useful reference work.

Materialist Film by Peter Gidal. $16.95. (Routledge, Chapman & Hall. Paperback. 189 pages. Index. Illus.)

According to its publisher, *Materialist Film* is "a polemical introduction to the avant-garde and experimental in film." Its author does not provide a definition of materialist film, does not understand the basic difference between a noun and a verb, and is incapable of expressing his ideas in such a manner as to be understandable to the general public. In view of the author's limitations, a reviewer is limited in his interpretation of the book and can only assume that it is intended for those who speak the same pseudolanguage as Mr. Gidal. God help us and the English language if these people ever get taken seriously.

May 1990

Fasten Your Seat Belts: The Passionate Life of Bette Davis by Lawrence J. Quirk. $21.95. (William Morrow. Hardcover. 464 pages. Index. Illus.)

Lawrence J. Quirk has produced a very thorough and lengthy (but never boring) biography of Bette Davis. It is a book which certainly reveals Miss Davis' wrinkles, but it also demonstrates an understanding of her undoubted talents both as an actress and, equally important, as a personality. Quirk has captured both the toughness and the vulnerability of a legendary lady. Along the way, he also discusses a number of "names" with whom Bette Davis has worked—his comments on Sanders are a revelation. Recommended.

Outfoxed by Alex Ben Block. $19.95. (St. Martin's Press. Hardcover. 304 pages. Index. Illus.)

Outfoxed is a fascinating account of the creation and formative years of the Fox Broadcasting Company. As with all good books, there is a villain—oilman Marvin Davis, who buys a half-share in 20th Century-Fox, with government support, for $116 million, and sells it six months later for $160 million. (Presumably the same people in government responsible for that deal are now overseeing the savings and loan industry.) There are heroes, in the unlikely persons of Barry Diller and Rupert Murdoch. And there is a tragic heroine—Joan Rivers. *Outfoxed* offers one of the best portraits of the comedienne that I have read. As presented here, she is both vulnerable and emotional, characteristics which seem far removed from the Rivers persona as seen on American television.

More and more of these serious, quasieconomic studies of the American media are being published, and they are welcome if disturbing records of the craving and in-fighting for power of this country's executives. They make Louis B. Mayer and Harry Cohn seem positively nice!

Mary Pickford: America's Sweetheart by Scott Eyman. $19.95. (Donald I. Fine. Hardcover. 342 pages. Index. Illus.)

I felt that Scott Eyman's biography of Mary Pickford started rather badly by discussing the subject's dreams, with which neither he nor any other writer could be familiar. That was the one weak part of the book—just a couple of opening paragraphs. All that followed is everything a first-rate biography should be, as Eyman discusses Pickford's career in detail and devotes about a quarter of the book to the star's postmovie life, the period which I found the most interesting and the most revealing. Eyman is not circumspect. He documents Pickford's alcoholism—"the family curse," as Lillian Gish once described it to me—and her often-contemptuous treatment of Buddy Rogers. Rogers did not cooperate with Eyman on the book, and one can understand why, although he is treated more than fairly in a biography which does not mention Rogers' unkept promise to Pickford not to remarry, or her fondness for cruelly comparing him to Douglas Fairbanks—"a real man."

Less hard to understand than Buddy Rogers' behavior is Eyman's apparent failure to use the voluminous Mary Pickford Collection of papers at the Academy of Motion Picture Arts and Sciences. David Pierce's filmography is well worth the price of the book.

Pickford had written her own autobiography and has been the subject of two biographies, both published in 1957, in which she was paired with Douglas Fairbanks. Those earlier works are minor in comparison with Scott Eyman's book. Well written with humor and a sincere regard for its subject, meticulously researched, it is a no-holds-barred but scrupulously fair study of a great American business-woman and one of the American cinema's greatest personalities. Highly Recommended.

Speaking of Silents: First Ladies of the Screen by William Drew. $24.95/$39.95. (Vestal Press. Paperback/Hardcover. 298 pages. Index. Illus.)

Speaking of Silents is a beautifully produced volume of narrative interviews with ten silent stars: Madge Bellamy, Eleanor Boardman, Leatrice Joy, Laura La Plante, May McAvoy, Patsy Ruth Miller, Colleen Moore, Esther Ralston, Blanche Sweet, and Lois Wilson. A basic filmography complements the book, and over 300 rare photos (and the majority I have never seen in print before) accompany the text.

It is a lovely book, but at the same time it is a disappointing one. There are no interviews here with reclusive stars of the silent era. All of the actresses included have been fairly accessible in terms of interviews; a number of the pieces have appeared in different formats in *American Classic Screen*; and four of the actresses have written their own autobiographies. There is, sadly, not too much new material here, and some of the interviews begin to read like catalogs of films, "then I did this, then I did that," etc. Finally, it is hard to understand, and no explanation is given, why the author does not bring his introductions on each subject up to date. Sad as it is, Bellamy, Joy, McAvoy, Moore, Sweet, and Wilson are dead. Why not say so? Why is Laura La Plante left living with a relative, when for some considerable time now she has been a resident of the Motion Picture Country House? Why is Patsy Ruth Miller's autobiography not mentioned? However, despite its faults, *Speaking of Silents* is a book which anyone who loves silent films cannot be without. Recommended with reservations.

The British Documentary Film Movement: 1926-1946 by Paul Swann. (Cambridge University Press. Hardcover. 216 pages. Index. Illus.)

Paul Swann provides a social and political history of the British documentary film from the founding of the Empire Marketing Board in 1926 through the demise of the Ministry of Information twenty years later. He gives John Grierson his due, but also attempts to emphasize the importance of Paul Rotha. (Swann is so determined to promote Rotha that he makes the mistake of claiming that the latter's 1930 book *The Film till Now* predates the 1926 publication of Terry Ramsaye's *A Million and One Nights.*) The book is meticulously researched and footnoted, and, in keeping with the British documentary, it is all rather dull.

Box-Office Buckaroos: The Cowboy Hero from the Wild West Show to the Silver Screen by Robert Heide and John Gilman. $19.95. (Abbeville Press. Paperback. 207 pages. Index. Illus.)

Box-Office Buckaroos is a revised edition of a 1982 book titled *Cowboy Collectibles.* As its name suggests, this volume is chockful of illustrations of all types of entertainment-related cowboy memorabilia, everything from Whitman's Little Books to Hopalong Cassidy earmuffs. Despite the title, *Box-Office Buckaroos* also includes information and illustrations on cowboy memorabilia unrelated to movies, radio, and television. It's a fun book and a must for anyone interested in collecting movie memorabilia. In that the authors do not indicate current values of any item, it will also never date (one of the reasons perhaps for its reprinting).

Unsold Television Pilots: 1955 through 1988 by Lee Goldberg. $45. (McFarland. Hardcover. 635 pages. Index.)

I get the impression that the television industry is going to be better documented than the film industry. This latest addition to the growing library of television reference works documents every unsold television pilot, television season by television season, broken down by category, and with credits and a brief summary of each plot. An invaluable index makes it possible to track down hard-to-find television credits for everyone from Peggy Ann Garner (featured in a 1962 pilot for *The Seekers*) to Steven Spielberg (director of a 1973 pilot titled *Savage*). It is interesting to learn that Lillian Gish was starred in two unsold pilots—in 1976 and 1978. In view of the television industry's supposed opposition to drug abuse, it is unfortunate to read in the introduction a comment by *Family Ties* producer Gary David Goldberg that he "might get stoned" and might look at one of his pilots. Recommended (the book that is, not getting stoned).

Pharmacy and Medicine on the Air by Mickey C. Smith. $22.50. (Scarecrow Press. Hardcover. 175 pages. Index. Illus.)

As I have noted before, it is amazing the topics which authors can think up. This slim volume documents reference to pharmacy on radio, with sections of pharmacy characters (such as *The Great Gildersleeve's* Richard Q. Peavey) and drug advertising. There are even statistical tables documenting which shows were sponsored by which drug companies between 1935 and 1955, and comparison tables enabling researchers to identify, for example, how many shows were sponsored by Bayer in 1954 compared to 1939. The detail is extraordinary, and the author (a professor of health care administration) demonstrates an entertaining style of writing. I hope somebody out there cares about all this, because he has undertaken a prodigious amount of research, and, just like old time-radio, he deserves an audience.

Ann Sothern: A Bio-Bibliography by Margie Schulz. $37.95. (Greenwood Press. Hardcover. 276 pages. Index. Illus.)

Margie Schultz has done a magnificent job of documenting Ann Sothern's career. Following a concise, yet detailed, biographical essay, Ms. Schultz provides documentation on Ann Sothern's career (both as Sothern and Harriette Lake) in films, theatre, radio, television, and on disc. (She even includes a listing of products which the actress has endorsed!) There is also a more-than-400 entry, annotated bibliography, which is quite extraordinary in its breadth and depth. What I like about this book is that its author never comes up with the obvious. In the entry on the Maine-located *The Whales of August,* she includes a review not from a major source, but from one of Maine's newspapers, which examines Sothern's performance from a regional viewpoint.

Many of the books in the Bio-Bibliography series from Greenwood are typeset. Unfortunately, this one is not. It has been printed directly from a typescript, and one can only wish that the typeface was more appealing, and that for some sections of the manuscript the typewriter ribbon had been replaced. Despite the "home-made" look of this rather expensive volume it is recommended.

Orson Welles: A Bio-Bibliography by Bret Wood. $39.95. (Greenwood Press. Hardcover. 377 pages. Index. Illus.)

Orson Welles: A Bio-Bibliography covers much the same territory for its subject as did the previous volume on Ann Sothern. Obviously, Welles is a somewhat more important film personality than Miss Sothern and so his volume is considerably lengthier. He doesn't warrant a typeset book, but he does get a cleaner typewriter ribbon. The Sothern book is compiled with fervor. The Welles book with a scholarly attention to detail. Recommended.

Andy Warhol: Film Factory edited by Michael O'Pray. $37.50. (Indiana University Press. Paperback. 192 pages. Index. Illus.)

Published by the British Film Institute and distributed in the U.S. by Indiana University Press, *Andy Warhol: Film Factory* (not *The Film Factory,* as the publisher's publicity calls it) collects a group of essays on Warhol's filmmaking activities, essays which vary in quality from the entertaining to the unreadable. Unfortunately, one learns a lot about what others think of Mr. Warhol and his films, but little about why Warhol made them or, for that matter, why he stopped making them. Of course, because of Warhol's fame (which has lasted far beyond its allotted 15 minutes), the films take on mythic proportions in this book. I can certainly remember being amused and possible slightly outraged by the films when they first appeared—the latter reaction because they were so badly made—but in 1990, some twenty years later, I can summon up little enthusiasm for a rescreening of them, and this book does *not* help change my mind. A basic filmography and bibliography are also included.

Directed by Vincente Minnelli by Stephen Harvey. $29.95. (The Museum of Modern Art/Harper & Row. Hardcover. 315 pages. Illus.)

Here is a beautifully conceived and produced tribute to the director, documenting each of his films in a scholarly yet readable fashion. Detailed credits are provided for each feature, and, needless to say, there is a breathless foreword by Liza Minnelli. Unfortunately, I cannot share Liza's enthusiasm for "daddy's pictures," which have always seemed to me beautiful but dull, rather like much of the work of Rex Ingram from an earlier period of film history. Gorgeous as is this volume, it should have contained an index, and it is surprising that the Museum of Modern Art should have permitted a book to appear under its aegis without such an obvious necessity.

Take One: Television Directors on Directing by Jack Kuney. $14.95/$39.95. (Praeger. Paperback/Hardcover. 191 pages. Index. Illus.)

Take One is the first book I have come across in which television directors are permitted to discuss their work in intelligent detail. The author introduces and interviews nine such gentlemen and one lady: Kirk Browning, Sandy Grossman, George Schaefer, Marc Daniels, Don Mischer, Paul Bogart, Lloyd Tweedy, Larry Auerbach, Emily Squires, and George Paul. The interviews cover just about every area of television production: sports, comedy, drama, children's programming, talk shows, and cultural events.

June 1990

Jennifer Jones: A Bio-Bibliography by Jeffrey L. Carrier. $35. (Greenwood Press. Hardcover. 155 pages. Index. Illus.)

As with the Alice Faye Bio-Bibliography, it is a little difficult to understand the need for a book such as this, intended as a research tool for those interested in an academic study of the career of Jennifer Jones. However, as was also true with the Alice Faye volume, this is a superior reference work. A concise biography is followed by a chronology, a filmography, a listing of radio, theatre, and television credits, a list of awards and nominations, and an annotated bibliography. The filmography includes not only complete credits for each film, but also a synopsis, a couple of quotes from contemporary reviews, a factual note, and, best of all, a first-rate listing of additional review and other sources. Carrier is easy to please, and he finds no fault with any of the entries in the bibliography. This reviewer is less easy to please, but he did like *Jennifer Jones: A Bio-Bibliography.*

Song & Dance: An Encyclopedia of Musicals by Sheryl Aumack. $29.95. (Sea-Maid Press. Paperback. 463 pages. Index.)

Quite obviously a labor of love, *Song & Dance* provides credit information for approximately 400 American-produced musicals. For each film, the author offers production and cast credits, a listing of the songs (with composers and lyricists), a two- or three-line synopsis, and a note as to whether the soundtrack is available on disc. The format for this spiralbound book is simple yet reliable. A series of indices provide references to the films by shows and performers, producers, directors, writers, and songs and songwriters. A useful reference work for anyone interested in the American musical film from 1933 to the present.

Hollywood Art: Art Direction in the Days of the Great Studios by Beverly Heisner. $39.95. (McFarland. Hardcover. 400 pages. Index. Illus.)

This is a valuable study of Hollywood art direction from the Twenties through the Fifties, in which the author equates the style of a studio's productions to the work of its art department. As a consequence, the work of art directors such as William Cameron Menzies, Anton Grot, Richard Day, and Hans Dreier is discussed in relationship to the studios for which they worked, with chapters on United Artists, M-G-M, Warner Bros., Paramount, 20th Century-Fox, RKO,

Columbia, and Universal. Two introductory chapters are devoted to the beginnings of art direction in the motion picture (unfortunately no new research is evident here) and art departments under the studio system.

It is, perhaps, a little difficult to understand the reasoning behind a chapter on United Artists, which was, after all, a distributor of films from various producers, each of whom must have had, and did have, different styles. Similarly, it is questionable whether Selznick International should be lumped with M-G-M, because of *Gone with the Wind*. In fact, these two questions bring up a general problem with the book in that some art directors—in this particular case William Cameron Menzies—overlap studios. Also, with regard to Menzies, I would have welcomed more discussion of his relationship to various directors, such as Sam Wood. It is very obvious that Wood's best work is such because of Menzies' involvement in the productions. I could also have wished for clarification of the art direction credit for *Camille* and *Salome,* which generally goes to Natacha Rambova, but which should perhaps belong to Harold Grieve (not mentioned in the text). There is at least one major mistake in that the author discusses *Becky Sharp* as an RKO production, which it is not—it was made for Pioneer Pictures and only released by RKO.

Despite such shortcomings, *Hollywood Art's* good points far outweigh its few negative aspects. High among the former is a comprehensive filmography of the work of more than 200 art directors. Recommended.

Film and Propaganda in America: A Documentary History, Volume I: World War I edited by Richard Wood. $75. (Greenwood Press. Hardcover. 502 pages. Index.)

The purpose of this volume is to gather together and reprint in their entirety various documents from the National Archives, relating to the government and the film industry between 1914 and 1918. It is an excellent idea, but the end result is disappointing. Many of the documents seem, quite frankly, to have no value. They are all reprinted on large format pages—8 1/2 x 11—resulting in an expensive book, a good third of which is blank paper. The compiler has made no effort to explain the relevance of the documents or to refer the reader to modern published essays or books which might help in an understanding of their value. A bibliographic essay is extremely incomplete and lacks a number of important film-related books and articles. What is included is sometimes treated in a cavalier fashion; Kevin Brownlow's *The War, the West and the Wilderness* is described thusly: "contains beautiful stills."

Four further volumes are planned in this series, under the editor-in-chiefship of David Culbert, concluding with documents on the Vietnam War.

The Encyclopedia of Hollywood by Scott Siegel and Barbara Siegel. $40. (Facts on File. Hardcover. 499 pages. Index. Illus.)

The Encyclopedia of Hollywood is a good idea which does not succeed. Its 700 entries cover a variety of topics, not only actors, actresses, directors, and other personnel, but also crafts, studios, and various general subjects (such as Animals in Films and Disaster Movies). It is good to find entries on subjects which often remain a puzzle to the general reader, subjects such as Deep Focus, Below the Line, or Answer Print. Quite obviously, for the casual reader, this book will do just fine, but for anyone with a knowledge of film history, it will be both a disappointment and an irritation. The entries are written in too simplistic a style; too many begin with generalities, describing an individual as "highly respected" or "influential," with the remainder of the entry failing adequately to explain why such a description fits. Notations as to published biographies or autobiographies would have considerably enhanced the book.

We read that Lillian Gish "returned to the screen with a splash" in *Duel in the Sun,* whereas she had returned to the screen without a splash some three years earlier in *The Commandos Strike at Dawn.* Clara Bow's career prior to 1925 is characterized as consisting of "small parts," hardly an accurate description of her role in the 1923 feature, *Down to the Sea in Ships.* Stock footage, according to the authors, "has rarely been used in recent decades," a statement which is totally untrue, as witness its use in hundreds of TV productions, the dozens of stock footage shots in the 1988 feature *Sing,* or even the just-released *Teenage Mutant Ninja Turtles.* Under the heading of "Directors," is a totally incomprehensible statement that "a director may be considered an auteur without writing, producing, or starring in a production."

What does that mean? I counted three factual errors in the entry on Blanche Sweet. And so on, and so on.

A Darling of the Twenties by Madge Bellamy. $19.95. (Vestal Press. Paperback. 208 pages. Index. Illus.)

I had met Madge Bellamy through the years on a number of occasions, but had never considered her a particularly interesting silent film star and never really felt much reason for talking with her in depth. The loss is mine, as evidenced by her splendidly open and honest autobiography, *A Darling of the Twenties,* which appeared this March, only a few weeks after her death. Without rancor, with surprising acceptance of the failings of both herself and those around her, Miss Bellamy has written a book which both fascinates and shocks with its candor. The epilogue, in which the actress documents her life in the past 45 years—the autobiography ends in 1945—is almost self-abusive in its speedy summation of years spent in abject poverty, in a life spent hiding her true identity. I could mention various incidents about which Madge Ballamy writes, but I don't want to spoil it for the reader. Suffice it to state that this book ends any suggestion that the silent film era was the age of innocence. *A Darling of the Twenties* also includes a collection of letters written to Madge Bellamy, an introduction by Kevin Brownlow, and a filmography compiled by William Drew. Highly Recommended.

July 1990

Directory of Blacks in the Performing Arts, Second Edition, by Edward C. Mapp. $57.50. (Scarecrow Press. Hardcover. 612 pages. Index. Illus.)

Some 1,100 Blacks in all areas of the performing arts are included in this excellent and unique reference work. Each entry includes a categorization, birth and death information, career data (often too brief), and listings for education, honors, club memberships, radio, television, theatre, and film work, musical compositions, and, in some instances, an address. There are omissions—for example, Carlton

Moss (*The Negro Soldier*) and Arthur Duncan (of *The Lawrence Welk Show*) are not included—and some entries are not current (for example that on the Nicholas Brothers). An index categorizes each individual by occupation from actor to ventriloquist (Willie Tyler), and also included is a bibliography and a useful directory of organizations. Recommended.

This Island Earth edited by Philip J. Riley. $19.95. (MagicImage Filmbooks. Paperback. Unpaged. Illus.)

Another entry in Riley's excellent series of reprints of Universal scripts, this volume is devoted to the 1954 science fiction feature. Aside from the annotated script of screenwriter Franklin Coen, the book also includes a commentary by Forrest J. Ackerman, a foreword by Coen, an introduction by leading man Jeff Morrow, a detailed account of the production by Robert Skotak, and a reprint of the original pressbook. As with other volumes in the series, I wish that some effort had been made to indicate differences between the published script and the released film. Recommended.

Cary Grant: A Bio-Bibliography by Beverly Bare Buehrer. $35. (Greenwood Press. Hardcover. 216 pages. Index. Illus.)

As with all books in the Bio-Bibliography series, *Cary Grant* includes a biography, chronology, filmography, discography, annotated bibliography, and a listing of appearances on radio, television, and the stage. This entry in the series is particularly well written, and includes considerable and valuable documentation. Recommended.

Maureen O'Sullivan: A Bio-Bibliography by Connie J. Billips. $35. (Greenwood Press. Hardcover. 224 pages. Index. Illus.)

This book contains all the sections noted in the previous review, together with reprints of two short stories by Miss O'Sullivan, which first appeared in *The Ladies' Home Journal.* This entry in the series is considerably poorer than earlier books. The filmography does not include synopses, commentary or sample reviews. A similar complaint can be made of the entries for Miss O'Sullivan's work on stage. The bibliography does include separate listings for film and theatre reviews, but instead of listing them by film or play, they are listed in some strange order, requiring the reader to look for reference numbers at the end of each film or play and then match those numbers to the bibliography—it is all very odd.

The Film Handbook by Geoff Andrew. $15.95. (G. K. Hall. Paperback. 362 pages. Index. Illus.)

This is one of a series of Performing Arts handbooks initially published in England by Longman. It consists of essays on some 200 directors, with each entry including basic factual information, suggestions as to other directors after whom the subject models his work, and a list of films most necessary for viewing. There is also a bibliography for each director, notable for its omissions, and a very shoddy group of listings in view of G. K. Hall's claim to be a supposed academic publisher. The book concludes with a glossary of genres and terms, another incomplete bibliography, and information on film magazines, institutions, and festivals, country by country. The best part of this book is the essays, which are entertaining, thoughtfully written and often provocative. Unfortunately, most of the information contained herein can be readily found in many other books, and it is difficult to understand the need for a work such as this.

Writers in Hollywood: 1915-1951 by Ian Hamilton. $25. (Edward Burlingame/Harper & Row. Hardcover. 326 pages. Index. Illus.)

If you want to read a scholarly history of screenwriting in Hollywood, Tom Stempel's *Framework* (1988) is highly recommended. If your desire is an anecdotal, entertaining history of screenwriting, in which accuracy is not too important, then *Writers in Hollywood* is for you. When the acknowledgements page describes Academy of Motion Picture Arts and Sciences' archivist Sam Gill as a staff member of the Production Code Administration Library, you know the author has problems with his facts. Throughout, author Hamilton quotes sources which are entertaining rather than reliable. For example, he states that Anita Loos' first movie credit was for *Macbeth,* co-shared with William

Shakespeare, and quotes Loos, "If I had asked they would have given me top billing." Very funny, but Anita Loos was not credited as the writer of *Macbeth.* Her husband-to-be, the much maligned John Emerson, was the acknowledged screenwriter. That is not a fact which is difficult to check. Similarly, it might be amusing or clever to claim that a listing of the credits of C. Gardner Sullivan "read like an extract from a soft-porn catalog," but is that really a very fair, or even intelligent, comment? This is a book with too much fun and not enough fact.

Gunsmoke: A Complete History and Analysis of the Legendary Broadcast Series by SuzAnne Barabas and Gabor Barabas. $45. (McFarland. Hardcover. 832 pages. Index. Illus.)

I feel guilty at devoting such little space to such a large book, but there is nothing to add to what the title states. This is the complete history of *Gunsmoke,* with a comprehensive episode-by-episode guide to both the radio and television series, together with much, much more. It's extraordinary, a superb work of documentation, formidable in its detail. What more can I say except highly recommended.

Meet Frank Capra: A Catalog of His Work compiled by the staff of the American Film Institute Catalog (Patricia King Hanson, executive editor; Alan Gevinson, associate editor). (Stanford Theatre Foundation/National Center for Film and Video Preservation. Paperback. 97 pages. Index.)

Published in connection with a tribute to Frank Capra in Palo Alto, California, and not available for sale, this pamphlet should be available by writing the AFI. In the same style as the *American Film Institute Catalog,* it documents all of the director's films, with complete credits, synopses, detailed notes, and listings of citations. Also included is an appendix documenting films which are sometimes attributed to Capra, but are actually the work of others. A superb piece of research and documentation, and highly recommended.

The Movie List Book: A Reference Guide to Film Themes, Settings, and Series by Richard B. Armstrong and Mary Willems Armstrong. $29.95. (McFarland. Hardcover. 377 pages.)

Here is a reference work which is both worthwhile and often futile. Under various headings, the authors provide a brief essay on the subject and a list of films or made-for-TV movies with dates. It is useful to find a ready reference source such as this for series (even British ones are included). Unfortunately, as regards themes and settings, it is remarkably incomplete. There is an entry for Military Academies, but none for schools, colleges, or universities. Apartments are included, but not houses. There is neither rhyme nor reason to the selection of subjects. It is almost as if the more obscure the subject, such as "Dates in Title" or "Earth's Core," the more likely it is that it will be included. Also, the essays often mention all the titles included in the accompanying lists, and it might have saved a lot of paper (and trees) simply to have included the dates with the titles in the essays and omitted the lists. A bibliography on each subject would also have been invaluable.

Major Film Directors of the American and British Cinema by Gene D. Phillips. $45. (Lehigh University Press. Hardcover. 290 pages. Index. Illus.)

This is a revised edition of Phillips' 1973 book, *The Movie Makers.* It contains succinct career evaluations on eight American directors: Charles Chaplin, Howard Hawks, Alfred Hitchcock, George Cukor, George Stevens, Fred Zinnemann, Stanley Kubrick, and Francis Coppola; and six British filmmakers: Carol Reed, David Lean, Joseph Losey, Bryan Forbes, John Schlesinger, and Ken Russell. Despite having interviewed quite a few of the directors, Phillips offers little new material, but each chapter is well thought out and provides a useful background to each man.

Scarlett's Women: Gone with the Wind and Its Female Fans by Helen Taylor. $12.95/$35. (Rutgers University Press. Paperback/Hardcover. 275 pages.)

Helen Taylor believes, quite rightly, that *Gone with the Wind* (both the book and the film) has had a tremendous influence on women. She asked various middle-class women in the United Kingdom to respond to a questionnaire with their experiences and views of the book and film. *Scarlett's Women* is the result of those responses, and it is a fascinating volume. She discusses the views of her women in relationship to Margaret Mitchell, the rape(s) of Scarlett, the racial issues raised by the book and the film, the ending, and so on. I am not a fan of *Gone with the Wind,* but I found this book thoroughly engrossing. Recommended.

August 1990

François Truffaut: Correspondence 1945-1984 edited by Gilles Jacob and Claude de Givray, translated by Gilbert Adair. $19.95/$50. (Farrar, Straus & Giroux. Paperback/Hardcover. 590 pages. Index. Illus.)

I read the published letters of director François Truffaut, sitting on my patio on a warm, summer's day, with the fountain playing—it's that kind of book. I had only a passing acquaintance with Truffaut, through my friendship with Jean and Dido Renoir, but as I read I could not but help recall another summer's day on that same patio, just after Truffaut's death, when my guests had been Dido Renoir and Truffaut's widow, Madeleine, and his daughter, Laura. And how, after dinner, we had screened *Sunrise,* because Truffaut had always wanted Madeleine and Laura to see it.

Just as I felt the emotion of remembrance, I felt a certain disappointment because, as with any collection of letters, the reader is unable to follow through on the writer's projects or comments. What was in the letters which he received to evoke such responses? Admittedly, a few letters to Truffaut are included, but they are not enough and often do not help towards an understanding of the topic under discussion. The editors have been meticulous in identifying all the individuals mentioned in the letters, but all too often they do not explain the point of the letter. It is also obvious that some of Truffaut's closest friends, such as Jean and Dido Renoir, were recipients of letters, copies of which were not kept, or else kept in touch by telephone, because there is nothing here addressed to them.

Film buffs will most certainly enjoy the considerable correspondence relating to Truffaut's interview book with Hitchcock—so much ego, conceit and flattery involved! Devotees of Truffaut may be disturbed, as I was, by a picture which emerges of a man not quite as likeable as he seemed in person, a conscienceless womanizer and user, given to easy flattery (as evidenced by his letters to Helen Scott and Annette Insdorf), an individual obsessed it seems with minutiae. And it is disturbing to note a certain homophobia in some of his letters.

Certainly, the price of the book is justified by a magnificent six-page letter to Jean-Luc Godard, in which Truffaut writes, "Like Sinatra, like Brando, you're nothing but a piece of shit on a pedestal." The animosity between the two men has long since dissipated, as evidenced by Godard's foreword in which he writes, "Francois is perhaps dead. I am perhaps alive. But then is there a difference." Highly Recommended.

Gone with the Wind on Film: A Complete Reference by Cynthia Marylee Molt. $45. (McFarland. Hardcover. 544 pages. Index.)

This is an extraordinary work of reference, containing everything (and in my case far more) than one would care to know about *Gone with the Wind.* All the characters in the book are described, together with the screen counterparts; characters in the book but not the film are also included. There are biographies for cast and crew members, together with listings of contenders for every role. Costumes and sets are described in astonishing detail. There are also listings for sheet music, collectibles, and every still photograph, with its code number and a description. An annotated bibliography includes not only all published references to *GWTW,* but also (and here it gets almost ludicrous in terms of what is included) articles, clippings, books, etc. relating to cast members, but which have no relevance to the film. Thus, you will find books on Marilyn Monroe listed, because they include references to *The Misfits,* in which Monroe's leading man was Clark Gable! The compiler has such a concern for detail that she includes both U. S. and foreign editions of books; for example, *Radie's World* has two entries for both the New York and London editions, although I am puzzled as to why the descriptions vary, because surely the book's content did not change from continent to continent. You can tell that I got so engrossed in this reference work that I was beginning to nit-pick. Recommended.

Live Television: The Golden Age of 1946-1958 in New York by Frank Sturcken. $29.95. (McFarland. Hardcover. 178 pages. Index. Illus.)

Frank Sturcken has produced a scholarly, heavily footnoted overview of the golden age of live television, with detailed accounts of the preeminent shows (*Kraft Television Theatre,* etc.) and the major figures (Pat Weaver, Fred Coe). This is not strictly speaking a reference book (although there is solid documentation in its pages), but more a readable survey of the period.

The Rock and Roll Movie Encyclopedia of the 1950s by Mark Thomas McGee. $25.95. (McFarland. Hardcover. 214 pages. Index. Illus.)

The Rock and Roll Movie Encyclopedia of the 1950s consists of essays on some 35 feature films, from *Because They're Young* to *Untamed Youth.* McGee comments on each film, and also provides cast and credit listings and a complete record of the music heard on the soundtrack. He also gives the reader a delightful summation of the film's initial reception, quoting both critical reviews and comments from exhibitors (culled from the pages of *Boxoffice* magazine). Two appendices provide information on which songs come from which films and who sang what in each. The subject of this volume holds not the slightest interest for me, but it says much for McGee's style and infectious enthusiasm that it held my attention, and I thoroughly enjoyed this trip through the downside of the Fifties.

A Deed of Death by Robert Giroux. $19.95. (Alfred A. Knopf. 275 pages. Index. Illus.)

Who killed William Desmond Taylor is a question to which no answer may be found. Certainly this new book by Robert Giroux offers a far more reasonable, and reasoned, solution to the crime than did Sidney Kirkpatrick's earlier, semifictional verdict. *A Deed of Death* is not only a first-rate account of the murder and its aftermath, but also a highly readable study of Hollywood filmmaking in the late 'teens. Highly Recommended.

Chinese Film Theory: A Guide to the New Era edited by George S. Semsel, Hou Jianping and Xia Hong. $39.95. (Praeger. Hardcover. 222 pages. Index.)

This is the first collection of essays on Chinese film theory to be published in English, gathering together the writings of a number of Chinese film scholars and theoreticians. It is perhaps surprising that there should even be any Chinese film theory, and, indeed, a casual reading of this book indicates there is little difference between theoretical thought in China and the United States. However, so little English-language material on Chinese cinema is available and any new source is welcome, particularly when, as in this book, there is enough historical documentation to make the theoretical discussion palatable.

Deaths of Noted Movie and TV Personalities compiled by Everett G. Jarvis. $19.95. (Wisdom for Youth Publishers. Paperback. 238 pages. Illus.)

This is a curious book. It contains death listings for some 2,400 film and television personalities, arranged chronologically, alphabetically and numerically by age and cause of death. In addition, it provides a statistical summary of deaths, a directory of cemeteries, and various other listings, some unrelated to the subject (such as the Oscar winners and nominees in the Best Picture, Actor, Actress, and Director categories). No explanation is provided as to the selection of names, and there are obviously omissions—no-one who died in the 1919 Spanish flu epidemic is listed—and some inclusions of individuals who are fairly obscure. In that *The International Film Necrology* (published by Garland in 1981) does not provide information on causes of death, this book does have value, and it is, in a rather weird way, a fun book through which to browse.

Garbo: Her Story by Antoni Gronowicz. $24.95. (Simon & Schuster. Hardcover. 476 pages. Index. Illus.)

Antoni Gronowicz met Garbo in 1938, and made love to her that first day (at least according to Gronowicz). He claims to have persuaded the actress to let him record her memories, for publication after her death. He made no tape recordings of these conversations, but, apparently, kept copious notes. Initially, the book alternates chapters by Gronowicz with chapters in the first person by Garbo, but is not too long before the book becomes Garbo alone.

There is a substantial difference between the style in which Garbo talks in the prologue and the manner in which she "writes." In the prologue, Garbo sounds suspiciously like bad dialogue from some of her films, with comments such as "The sun is shining the same way on flowers and on mud" or "A man gets married because he is in love; a woman pushes herself into love because she wants to get married." Later, the actress appears to have adopted a more restrained and atypical biographical approach to her career, noting the dates when her films opened in New York, at which theatres, and even quoting from *Variety* or *The New York Times.* (I really find it hard to believe that Garbo could care less what Mordaunt Hall thought of her films.)

While it is obvious that Gronowicz and Garbo were acquainted, it is not totally believable that the style of the book is not more the former's work. Neither Garbo nor her estate authorized publication. There are surprises and shocks in abundance here. Garbo talks of her lesbian relationship with Marie Dressler, whom she claims "taught me not to be ashamed of this kind of love," and goes on to detail similar couplings with actresses Dorothy Sebastian, Barbara Kent, and others. She is extremely negative in her appraisals of Leopold Stokowski (only after her money and unable to make love) and Gayelord Hauser, both of whom were supposedly close friends.

A considerable amount of the book is devoted to Garbo's relationship with Mauritz Stiller, whom she calls Moje. Despite all evidence to the contrary, she makes no reference to his being homosexual, indeed quite the opposite, documenting his affair with her and other women.

There is little question that *Garbo: A Biography* is destined to be a best-seller, a volume which will arouse controversy and be the subject of much discussion among film buffs. There is no point in recommending or not recommending it; every film buff is going to want this book, regardless of what I write.

Variety Broadcast-Video Sourcebook 1, 1989-1990 compiled by Marilyn J. Matelski and David O. Thomas. $14.95. (Focal Press. Paperback. 124 pages.)

Here is what should be an extremely valuable resource, a book which gathers together from the pages of *Variety* a multitude of documentation on television, radio, CATV, and home video. Also included is a brief history of *Variety*, a chronology of major events in the history of popular entertainment, a 1989 necrology, and a glossary of terms utilized by *Variety*. Unfortunately, not included is an index, which makes the material of only slightly more value than had it remained within the pages of weekly *Variety*, and which hardly helps as is the intention of the compilers, "to expand *Variety's* vast audience to include more students and teachers of higher education." A *Variety Sourcebook for Film-Theatre* is announced, and presumably this is only the first of an annual overview of the broadcast-video field.

Bugs Bunny: Fifty Years and Only One Grey Hare by Joe Adamson. $35. (A Donald Hutter Book/Henry Holt and Company. Hardcover. 192 pages. Illus.)

It's Bugs Bunny's 50th anniversary and doubtless, much to the chagrin of Daffy Duck, the event is celebrated with this sumptuous volume, replete with black-and-white and color illustrations and a sometimes irritatingly flippant but always factually accurate text. *Bugs Bunny: Fifty Years and Only One Grey Hare* includes prefaces by his two remaining living animators, Chuck Jones and Friz Freleng, a detailed filmography, and a particularly interesting section which successfully illustrates Bugs Bunny's origins in the films of Charlie Chaplin and other silent comedians. Recommended.

Tom Mix: The Formative Years by Paul E. Mix. $5.95. (PM Publications. Paperback. 48 pages. Illus.)

Paul E. Mix has written a short (37-page) biographical novel on Tom Mix's early years up through 1898 and appended to it a series of comments. Presumably if you like the first 37 pages of Mix's life, you will ask the author to continue the story—he is willing and able.

The White Brothers: Jack, Jules & Sam White interviewed by David N. Bruskin. $49.50. (Scarecrow Press. Hardcover. 531 pages. Index. Illus.)

In this volume, No. 10 in the Directors Guild of America Oral History Series, three brothers whose names are synonymous with the history of comedy shorts discuss their work, as a group and then individually. This is an important book, and yet I found it a little disappointing. I cannot explain why, except that at times the questions seemed naive or even pointless (relating as some of them do to factual matters well recorded in contemporary sources). It is hard to feel a rapport with an interviewer who appears to know Kenneth Macgowan only as the author of a book titled *Behind the Screen.* And I suppose the problem is that simply I did not like the interviewer's style. It may well be that the fault is mine and not his, but that's the way it is.

September 1990

Those Fabulous Serial Heroines: Their Lives and Films by Buck Rainey. $59.50. (Scarecrow Press. Hardcover. 537 pages. Index. Illus.)

Buck Rainey has compiled a very useful reference work, documenting the careers of some 46 leading ladies of both silent and sound serials. Each entry includes an introductory essay and a detailed filmography of the work of the actress both in serials and other genres. The entries are divided into two groupings—"Serial Heroines Nonpareil" and "Fearless Beauties in Distress"—with the first category including the major stars who had "true grit" and the latter covering "the second echelon." (Perhaps a little unfair a description, because two of the ladies here, Kathlyn Williams and Mary Fuller, were major stars, even if their serial contributions were less than those of Kay

Aldridge and Noel Neill, who are included in the first grouping and who are, in all honesty, virtually unknown outside of the serial genre.)

I was at first concerned that the book might give too much prominence to the leading ladies of sound serials, but such is not the case, and it is very nice to see Pearl While, Ruth Roland, Anna Little (a lovely lady), and others receiving their due. Unfortunately, no-one from the silent era is around to perform the task, and so introductions are contributed by Jean Rogers and Carol Forman. I must confess that I had not heard of either, but Rainey's description of the career of Miss Forman encourages me to seek out her work on screen. Recommended.

Madcap: The Life of Preston Sturges by Donald Spoto. $19.95. (Little Brown. Hardcover. 301 pages. Index. Illus.)

As one would expect from Donald Spoto, *Madcap* is an engaging and well-researched study of the life and career of writer-director Preston Sturges. Unfortunately, however, it provides no new information and basically covers the same territory covered by James Curtis in his 1982 biography, *Between Flops* (currently available in paperback from Limelight).

So You Want To Make Movies by Sidney Pink. $18.95. (Pineapple Press. Hardcover. 284 pages. Illus.)

Sid Pink has enjoyed a fascinating career as an independent producer from a school of filmmaking which no longer exists today. He worked as a production manager for Grand National in the mid-Thirties. He was the industry's first budget manager at Columbia from 1937-1938, until he proved himself Harry Cohn's equal in invective and was promptly fired. After a decade-long hiatus from the industry (10 years which were not without interest), Pink returned as associate producer to Arch Oboler on *The Twonky* and *Bwana Devil* in 1950-1951. He began his career as the man in charge in 1959, and for the next 11 years turned out an assortment of features in Denmark, Spain, and Puerto Rico, among which are *The Angry Red Planet, Reptilicus, Pyro, Madigan's Millions,* and *The Man from O.R.G.Y.*

There can be little question that Sid Pink was one hell of an entrepreneur. He is also one hell of a good storyteller. You may not always agree with the upbeat assessments of some of his films, but there is no question that Pink was (and probably still could be) a consummate filmmaker. He commences his story be telling of a series of lectures he attended at UCLA on the art and craft of filmmaking. Quickly he determined that was no way to learn about the industry. Like so many pioneers, he went out and he did it. His story should serve as a lesson for all would-be filmmakers. Highly Recommended.

Stages: Norman Lloyd interviewed by Francine Parker. $32.50. (Scarecrow Press. Hardcover. 296 pages. Index. Illus.)

Norman Lloyd has had the good fortune to be involved with so many legendary figures—Orson Welles, Charlie Chaplin, Jean Renoir, Lewis Milestone, Alfred Hitchcock, etc.—in the film industry. Equally lucky are those privileged to know Norman and to be able to enjoy his marvelous stories relating to both the stage and screen. He is a magnificent raconteur, and as he tells a story with that wonderfully authoritative voice of his, one feels a closeness with the best aspects of Hollywood history.

Happily, some of Norman's stories have been gathered together in this Directors Guild of America Oral History volume, along with narrative on his career as an actor, director, and producer. There are a few questions by Francine Parker, interspersed in the text, but it is mainly Norman Lloyd. In fact, it is more autobiography than an interview book.

It is a delightful and informative volume, so full of good humor and sound critical appraisal. I loved Norman's comment on the industry: "In our business, executives are convicted of embezzlement and then made heads of studios again, as if nothing had happened. The moral texture of this country must be affected; the young must conclude that this is the way to succeed." I was pleased to reencounter the hilarious tale of Bertold Brecht's reaction to Charles Laughton's playing with himself on stage during the run of *Galileo.* I laughed as much this time as I did the first time Norman told it to me in the kitchen of Jean and Dido Renoir. I am disappointed that not all of Norman's stories are included, but at the same time I am pleased because surely that means there is an obvious opportunity for a second volume. Highly Recommended.

Ava Gardner: A Bio-Bibliography by Karin J. Fowler. $37.95. (Greenwood Press. Hardcover. 256 pages. Index. Illus.)

Ava Gardner is a typical entry in the Bio-Bibliography series, of which it is No. 14. The book includes a biographical essay, a detailed inventory of films, with credits, synopses, and review annotations; listings of Ms. Gardner's work in television and radio; a record of soundtracks from her films; and a bibliography. I felt the biographical section could have been improved with some editorial help. At least two of the annotations in the bibliographic section are faulty, and there is a curious error in the synopsis of *On the Beach,* which states that only a handful of people have survived the total nuclear destruction of the earth, when surely it is the entire population of Australia which has survived, albeit temporarily.

Apropos of nothing, I recall as a very young man sitting in a London bar and watching Ava Gardner arrive. I thought she was the most beautiful woman I had ever seen, and my memory still tells me that she was.

Renoir on Renoir translated by Carol Volk. $14.95/$47.50. Cambridge University Press. Paperback/Hardcover. 421 pages. Index. Illus.)

There are a few times when I get genuine pleasure upon receiving a new book. The day *Renoir on Renoir* arrived is just such an occasion. The book gathers together various French-language interviews with Jean Renoir, primarily from Cahiers du Cinéma and also from a couple of French television programs. In great detail, and sometimes with humor, the great director discusses his various films, and answers, with I believe a certain amount of self-satisfaction, the always respectful and often adulatory questions from the likes of François Truffaut and Jacques Rivette. His memory is phenomenal, and his understanding of both the art and business of filmmaking is revelatory. I only wish the book had been longer and had included interviews with Renoir from English-language publications. In that Jean Renoir spent the last years of his life, when he had little to do but give interviews, in the

United States, I suspect there is much more material available than is presented here. Highly Recommended.

As Thousands Cheer: The Life of Irving Berlin by Laurence Bergreen. $24.95. (Viking. Hardcover. 658 pages. Index. Illus.)

As Thousands Cheer is a monumental biography of a complex and basically not very nice human being who was born Israel Beilin and who as Irving Berlin became the greatest songwriter in America and probably the world. In reading the life story of the man who wrote "God Bless America" and "White Christmas," it is difficult to understand why he should have been such a great songwriter. I don't agree with the comment by one of the minor characters in the book that "It makes you believe in God, to think HE could create someone like Irving Berlin." More, it makes you believe in the intangible mystery of life.

As Thousands Cheer is meticulously researched, highly readable, and includes a 14-page chronology of songs written by Irving Berlin, from "After the Honeymoon" to "You Need a Hobby." Perhaps not the most memorable songs in his repertoire, but don't forget that in between there is "Alexander's Rag Time Band," "Always," "All Alone," "Blue Skies," "Puttin' on the Ritz," "How Deep is the Ocean," "Cheek to Cheek," "Easter Parade," "There's No Business Like Show Business," and so on and so on and so on. . . .

Japanese Films: A Filmography and Commentary, 1921-1989 by Beverly Bare Buehrer. $39.95. (McFarland. Hardcover. 360 pages. Index. Illus.)

The title is a little misleading, because, in reality, this is a study of only 80 or so Japanese films, but what is here is good. For each film, the author provides credits, a lengthy description of the feature (far more than a synopsis) and a critical study/historical overview. The book also includes a list of video and film sources, an annotated bibliography and an extremely useful glossary of Japanese film terms. This is not a major work on Japanese cinema, but it is, nevertheless, a useful book for anyone interested in the subject.

The Continental Actress: European Film Stars of the Postwar Era by Kerry Segrave and Linda Martin. $35. (McFarland. Hardcover. 304 pages. Index. Illus.)

Similar in format to the same authors' *The Post-Feminist Hollywood Actress* (published last year), *The Continental Actress* provides career surveys and filmographies (film titles and dates only) for 12 Italian, 2 Greek, 15 French, 6 German, and 6 Scandinavian actresses. In that it is difficult to find readily available information on many of the ladies included here, *The Continental Actress* is a welcome addition to the film reference bookshelf.

Laurel & Hardy: A Bio-Bibliography by Wes D. Gehring. $49.95. (Greenwood Press. Hardcover. 328 pages. Index. Illus.)

Like Mr. Laurel and Mr. Hardy, Wes D. Gehring is a pretty unusual character. He is an academic—a professor at Ball State University—who manages to write serious books on comedic subjects, and he accomplishes his task with humor and an obvious affection for those of whom he writes. He is also a poet, and two of his verses are included here—and unlike most modern verse, you actually enjoy reading them. Gehring's latest work includes a lengthy biography, an analysis of the screen characters of Laurel and Hardy, some contemporary reflections on the pair, and a biographical essay which will, I am sure, please John McCabe but not be to Scott Allen Nollen's liking. There is also a very detailed chronology, which covers not only Laurel and Hardy's live and careers, but also notes events in the lives of those around them. The book concludes with a filmography and a selected discography.

Laurel & Hardy: A Bio-Bibliography provides the researcher and the student with everything he or she needs to know about the comedy duo, and, quite frankly, it supercedes a lot of the books which have gone before. Recommended.

The Modern Horror Film by John McCarty. $15.95. (Citadel Press. Paperback. 244 pages. Illus.)

The Modern Horror Film is a "Films of" book covering fifty "recent" features from *The Curse of Frankenstein* in 1957 through *The Lair of the White Worm* in 1988. For each film, McCarty has provided detailed credits, a complementary essay and lots of (often gory) illustrations. John McCarty is such an entertaining and authoritative writer that he can win over even a disinterested reader such as myself. In fact, I am not totally disinterested, because the book does include one of the great horror films of all time, *Curse of the Demon,* written by the much underrated Charles Bennett. McCarty also quotes what is the most outrageous dialogue I have heard in a recent film (excluding Peter Greenaway's *The Cook, The Thief, His Wife and Her Lover*), and that is from Ken Russell's *The Lair of the White Worm:*

His question is "Do you have children?" Her response is "Only when there are no men about."

Robert Ryan: A Biography and Critical Filmography by Franklin Jarlett. $29.95. (McFarland. Hardcover. 308 pages. Index. Illus.)

Franklin Jarlett has compiled a sympathetic and extensively researched biography of Robert Ryan, whose performances have always exhibited a sensitivity sometimes at odds with the films in which he was appearing. Aside from a first-rate biography, the book includes a highly detailed filmography with critical commentary and review excerpts for each film, plus a listing of stage and television performances and other material. Recommended.

October 1990

Lina: DeMilles' Godless Girl by Lina Basquette. $22.95. (Denlinger's Publishers. Hardcover. 450 pages. Illus.)

Lina Basquette was an actress who enjoyed a career on screen which extended, with breaks, from 1916 through 1943. Her autobiography does not tell one too much about the films in which she appeared—only two, Lois Weber's *Shoes* and Cecil B. DeMille's *The Godless Girl,* are discussed in any detail. The book does, however, tell you a great deal about a woman who quite obviously never knew when to say No. With unbelievable candor, Miss Basquette reveals the intimate details of her childhood and her quintessential stage mother, her marriage to Sam Warner of the Warner Bros., her widowhood and the abusive treatment she received from Sam's brothers (including the eventual loss of her daughter), her second marriage to cameraman J. Peverill Marley, her love affairs with just about everyone from mobster Johnny Roselli to singer Nelson Eddy, her third marriage to British actor Henry Mollison (who would be horrified if he were still alive, or perhaps he would be delighted, at the amount of space devoted to the size, substantial, of his penis), and her attempted rape by Adolph Hitler.

It is all here—frankly and unabashedly revealed and mainly recounted through remembered conversations. Despite, or perhaps because of, it all, Miss Basquette seems to have come out of it OK. She breeds Great Danes—and anyone who loves dogs cannot be too bad. The book concludes with a short epilogue, provided by the editor, following up on the actress' life following her eviction from Nazi Germany. I hope this means there is another volume to come, because what little is revealed of Miss Basquette's later years makes me long to read more.

Lina Basquette has been the subject of a *New Yorker* profile by Barry Paris (the man responsible for the recent biography of Louise Brooks). I hope this does not mean there is going to be an attempt to turn the lady into another Miss Brooks. First of all, Lina Basquette is too much fun. For another, she is far more attractive in old age than was Louise Brooks, and, looking at early photographs of her, I am reminded more of the perkiness of Clara Bow than the cold beauty of Miss Brooks. In fact, in *The Godless Girl,* a rather bad film, Lina Basquette is a rather good actress.

As *Lina* reveals, Miss Basquette is a survivor. You can take her child away. You can threaten her with physical violence. But she is never down, never out.

P. S. If any reader was wondering what can be said for sexual intercourse in an unheated swimming pool, after experiencing it with Lawrence Tibbett, Lina Basquette reveals that it is clean.

Highly Recommended.

Camera Politica: The Politics and Ideology of Contemporary Hollywood Film by Michael Ryan and Douglas Kellner. $14.95. (Indiana University Press. Paperback. Index. Illus.)

Originally published in hardcover in 1988, *Camera Politics* is a readable, but decidedly Leftist, view of Hollywood films in relation to contemporary events and American society, from the mid-Sixties through the mid-Eighties. Chapters deal with all the usual subjects one would expect in a book such a this, the politics of sexuality (I can't help thinking Lina Basquette could have told the authors plenty on that subject), alienation and rebellion, Black radicalism, and the like. I rather enjoyed the discussion of Francis Ford Coppola and the quote describing him as the "Bay Area Mussolini."

Film and Propaganda in America: A Documentary History: Vol. II: World War II: Part 1 edited by David Culbert. $75. (Greenwood Press. Hardcover. 544 pages.)

I was more than a little critical of the first volume in this series. I am happy to report the second volume more than makes up for any failing earlier reported. This present book is devoted entirely to reprints of reports, many of them originally confidential, investigating the production of training films in Hollywood between 1940 and 1943. The documentation is fascinating, particularly the comments by independent producers of industrial and documentary films, critical of the costs of such productions by the Hollywood studios. The implication is that the majors were guilty of profiteering, but the reality is they were merely guilty of stupidity and mismangement. What is particularly valuable about this collection is the factual information, unrelated to the subject at hand, buried in its pages. For example, thanks to this book, I now know, from the testimony of its president, that DeLuxe Laboratories was founded in 1915 and incorporated in 1922.

Russ Meyer—The Life and Films by David K. Frasier. $39.95. (McFarland. Hardcover. 236 pages. Index. Illus.)

Russ Meyer—The Life and Films contains a brief biography and a detailed filmography, but consists primarily of almost 200 pages of detailed bibliography. Surprisingly, there is a considerable amount written about this extraordinary purveyor of sexual exploitation, who made one film of brilliance, *Beyond the Valley of the Dolls,* written by critic Roger Ebert, whom I was interested to learn has also written other films for Meyer under the pseudonym, "R. Hyde." Works of particular merit are identified with an asterisk, and one of the first I encountered appears by the entry for an article titled "Tits 'R' Russ." Hmm. The book is adequately illustrated with photographs of young starlets baring considerable bosoms which defy gravity.

Lawmen in Scarlet: An Annotated Guide to Royal Canadian Mounted Police in Print and Performance by Bernard A. Drew. $30. (Scarecrow Press. Hardcover. 296 pages. Index. Illus.)

A useful reference work, if of somewhat limited value, *Lawmen in Scarlet* provides entries for more than 200 films, 500 fictional works, and assorted television programs, radio shows, and theatrical ventures which have featured the men of the Royal Canadian Mounted Police. Fictional works are listed under authors, for whom brief biographies are provided, and film entries include credits and brief synopses. Any book which contains an entry for my favorite childhood television series, *Sergeant Preston of the Yukon,* featuring Yukon King, the world's stupidest dog, has a place on my bookshelf.

How Scripts Are Made by Inga Karetnikova. $26.95/$16.95. (Southern Illinois University Press. Hardcover/Paperback. Index.)

According to Eleanor Breese on the cover of this book, "We have needed a practical, readable textbook like this for a long time." Unfortunately, I have no idea who Eleanor Breese is, nor who are the "we" to which she refers. *How Scripts Are Made* probably will serve as a textbook for those who believe that scriptwriting can be taught by educators who have never had a script filmed. Perhaps the author is such a person, for she displays little practical knowledge of the filmmaking process, and I sincerely doubt that what she publishes as the scripts for *A Trip to the Moon* and *The Great Train Robbery* are really such. I suspect that if the author was a working member of the film industry, she would not have written this book, but rather would have encouraged would-be scriptwriters to obtain a sample script (any sample script) and go to it.

Jean Arthur: A Bio-Bibliography by Arthur Pierce and Douglas Swarthout. $39.95. (Greenwood Press. Hardcover. 288 pages. Index. Illus.)

Jean Arthur, No. 15 in the Bio-Bibliography series from Greenwood, receives high marks from me right away for answering any potential complaints I might have as to structure or inclusionary criteria with a well-considered preface. Following a lengthy and informative biography, the authors provide a chronology, a filmography, listings of stage, television and radio appearances, a discography, a bibliography, and reprints of three magazine articles, including one each by John Wayne and Jean Arthur. While the authors fail to provide a total understanding of Jean Arthur's temperament and personality (something which is probably impossible in view of the lady's reclusivity and recent illness), they have provided a first-rate study of, and complete documentation on, her career.

Preston Sturges by Preston Sturges adapted and edited by Sandy Sturges. $22.95. (Simon & Schuster. Hardcover. 352 pages. Index. Illus.)

Preston Sturges was a brilliant writer and director, with more than a streak of what is perhaps best described as mild-mannered lunacy to his films and his life. A fair amount of that joyful insanity comes across in this book which is based, in part, on Sturges' unpublished autobiography and in part on other materials, presumably correspondence. It is an entertaining book, well worth reading in tandem with James Curtis' 1982 biography, *Between Flops.* I only wish either the publisher or the editor had seen fit to indicate the sources for the material. In that the autobiography covering Sturges' early life and work in the theatre takes up more than two thirds of the book, it would be nice to know the origins of the remaining third.

Movie Characters of Leading Performers of the Sound Era by Robert Anthony Nowlan, Jr. and Gwendolyn Wright Nowlan. $47.50. (ALA Books. Paperback. 398 pages.)

More than 400 players are included in this book, with each entry including a brief biographical sketch and a listing of roles (with descriptions) played. This is not an original idea, having been tackled by Tom Scott Cadden in *What a Bunch of Characters!* (1984), but the latter volume included only 50 stars. The main problem with the current book is that there is no way to retrieve information based on a knowledge of the screen character only. To do that, the reader will need to have copies of *Memorable Film Characters* by Susan Lieberman and Frances Cable (1984) and *Who Played Who on the Screen* by Roy Pickard (1988). However, as far as it goes, *Memorable Characters of Leading Performers of the Sound Era* is a highly detailed reference work and appears to be reliable. Certainly, a reading of the descriptions of an actor's roles gives a vivid indication of that individual's capability and range.

The Home Cinema: Classic Home Movie Projectors, 1922-1940 by Gerald McKee. (Gerald McKee. Paperback. 121 pages. Illus.)

Gerald McKee has done an excellent job of gathering together esoteric documentation on all manner of home movie projectors—8mm, 9.5mm, 16mm, 17.5mm, and 35mm—in use in the United Kingdom between 1922 and 1940. The history of each model, many of which I am sure are unknown to most readers, is told in detail, and there are illustrations of the projectors and reproductions of contemporary advertising. Appendices discuss the reasons why anyone would want to collect such projectors and some of the problems involved in their restoration, literature on the subject, and a listing of original projector lamp types.

The Home Cinema is available direct from the author, via airmail, for 14.16 pounds. The first edition is limited to 350 copies, and I would advise any interested reader to run, not walk, to your local bank for an international money order.

November 1990

Indiscretions: Avant-Garde Film, Video & Feminism by Patricia Mellencamp. $29.95/$12.95. (Indiana University Press. Hardcover/Paperback. 256 pages. Index.)

Indiscretions is an academic history of the American avant-garde film from the 1960s to the present. The author adopts a feminist viewpoint of her subject and assumes, in a rather superior academic fashion, that the reader has a familiarity with the theoreticians which she quotes ad nauseam. It is amazing how many individuals, unknown to more than 99.9% of the population, have gained legendary status as far as a minuscule academic film community is concerned. However, despite my criticism, *Indiscretions* is readable, and both irritates and fascinates, for which the author deserves credit.

Patricia Mellencamp makes a very interesting point that "Film history so far has largely been left to men, who have paid scant heed to the gendered bias of its writing." Unfortunately, she then loses all my sympathy by writing of Mary Pickford that she "became a very powerful woman—only to have the variables of age and stereotype force her into retirement, to be mocked in *Singin' in the Rain*." What nonsense, and so easily disputed by a viewing of any of Pickford's sound features.

TV and Studio Cast Musicals on Record: A Discography of Television Musicals and Studio Recordings of Stage and Film Musicals by Richard Chigley Lynch. $49.95. (Greenwood Press. Hardcover. 344 pages. Index.)

This is the third reference discography by Richard Chigley Lynch and published by Greenwood Press. It includes 657 entries, for each of which Lynch provides lyricist and composer credits, cast and song listings, disc number, and either date of original telecast or year of recording (if a studio recording). In case you were wondering, as did I when the book first arrived, what is the latter, apparently, it is a recording of a show by other than the original cast. For example, the entry for *Thoroughly Modern Millie* is for a British recording of the songs from the 1967 film by Diana Dors, David Kernan and Millicent Martin. I quite enjoyed browsing through this volume. Does anyone remember a 1956 recording called *My Square Laddie*, which appears to be a parody of *My Fair Lady*, with Reginald Gardiner, ZaSu Pitts and Nancy Walker? Among the included songs are "I'm Kinda Partial to His Puss" and "I Could Have Boozed All Night."

Also included in the book is a chronology, a performer index, and an index for everyone else. An excellent work of reference, and recommended.

TV Movies and Video Guide, 1991 Edition by Leonard Maltin. $6.95. (Signet Books. Paperback. 1,330 pages.)

The new edition of Leonard Maltin's *TV Movies and Video Guide* continues the same reliable format adopted over the past 20 years. For each film, director and leading player credits are provided, along with year of release, running time, a few-lines commentary, and rating of between zero and four stars. Films available on videotape are so identified, while made-for-television movies are designated as below average, average, or above average. A particularly useful addition this year is a listing of mail order sources for videotapes and discs. Even the introduction is entertaining and informative—Leonard Maltin confirms unequivocally that Daniel Day-Lewis does have a hyphen in his name.

I never thought I would agree with Larry King, but I can only reiterate his comment, "It belongs next to every TV and VCR in every home." Highly recommended.

Acting in the Cinema by James Naremore. $12.95. (University of California Press. Paperback. 316 pages. Index. Illus.)

First published in hardback in 1988 and now available in paperback, *Acting in the Cinema* is an interesting and informative study with emphasis on a select group of performances: Lillian Gish in *True Heart Susie,* Charles Chaplin in *The Gold Rush,* Marlene Dietrich in *Morocco,* James Cagney in *Angels with Dirty Faces,* Katharine Hepburn in *Holiday,* Marlon Brando in *On the Waterfront,* and Cary Grant in *North by Northwest.* The style is somewhat pedantic, but occasionally enlivened by interesting tidbits, such as the reprinting of diagrams from Charles Aubert's 1927 text, *The Art of Pantomime.* Perhaps Naremore tends to concentrate a little too much on well-known films, and when he does delve into film history he is on shaky ground, such as identifying 1916 as the first year in which celebrity characters make their way into Hollywood fiction—there are earlier examples of actors and actresses appearing as themselves on screen.

Picture Personalities: The Emergence of the Star System in America by Richard deCordova. $26.95. (University of Illinois Press. Hardcover. 160 pages. Index. Illus.)

Picture Personalities is an interesting study of the creation of the star system in the American film industry up through the early Twenties, based in large part on primary sources. I was particularly impressed that the author had taken the trouble not only to discover the existence of, but also to examine, Florence Lawrence's fan mail in the Los Angeles County Museum of Natural History. The index is somewhat incomplete, and do not be put off by the introduction, which is rather tedious and brings up the same tired, old academic names who have bored us all for years (as Noel Coward might have said—in fact did). This is a short book and there is a lot more to be said on the subject, but as far as it goes, *Picture Personalities* is thoroughly worthwhile. Recommended.

Hollywood Goes to War by Clayton R. Koppes and Gregory D. Black. $12.95. (University of California Press. Paperback. 384 pages. Index. Illus.)

Published in hardcover in 1987, the new paperback edition of *Hollywood Goes to War* is a welcome addition to the bookshelf. Through the examination of primary sources, such as the files of the Production Code Administration and the records of the Office of War Information at the National Archives, the authors have compiled a first-rate examination of the American film industry's contribution to the war effort. The films, the personalities, and the politics are all discussed in detail. I do not believe there has been a better book on the films of the Second World War, and I doubt there will be. Recommended.

Fred Allen's Radio Comedy by Alan Havig. (Temple University Press. Hardcover. 287 pages. Index. Illus.)

Fred Allen is enjoying something of a revival—and about time. A first-rate autobiography by Robert Taylor, *Fred Allen: His Life and Wit* was published last year, and now we have a volume covering Allen's radio comedy. Chapters are devoted to Allen as a satirist, *Allen's Alley* (1942-1949), the creation of radio comedy, and Allen's problems with network and advertiser censorship. The author takes an academic approach to his subject, but has obviously either read the scripts of, or heard all of, the Allen shows and enjoyed them and carefully footnotes his work.

Reading of the characters in *Allen's Alley,* it is disturbing to note that they and he are accused of racism. Minerva Pious' Mrs. Nussbaum was attacked by one listener in 1948 as "a grievous racial stereotype in a age where too many graves attest to the evils of racial stereotypes." And what would Americans today make of Senator Beauregard Claghorn, who swore off the movies after seeing a film in which Ann Sothern married Cary Grant—"I couldn't stand to see another Grant takin' anything Sothern."

Order in the Universe: The Films of John Carpenter by Robert C. Cumbow. $29.95. (Scarecrow Press. Hardcover. 240 pages. Index. Illus.)

Order in the Universe, No. 23 in the Filmmakers series, is the first book-length study of the films of John Carpenter from *The Resurrection of Broncho Billy* in 1970 to *Halloween 5: The Revenge of Michael Myers* in 1989. It includes a detailed filmography.

Space Funnies. $5.95. (Archival Photography. Paperback. 32 pages. Illus.)

Robert Brosch's Archival Photography has devised a process for reprinting vintage comic books, and the first to be offered is Basil Wolverton's *Space Funnies* (with three episodes of Spacehawk). Inside the back cover is an amusingly written history of the original series. I know nothing about comic books, but, as a casual observer I can now report that the reproduction is excellent, and, of course, the paper being used is of a far higher quality than that on which the original comics were printed. I am sure the series will prove popular.

December 1990

Emile Cohl, Caricature, and Film by Donald Crafton. $67.50. (Princeton University Press. Hardcover. 404 pages. Index. Illus.)

For the animation buff, *Emile Cohl, Caricature, and Film* is the perfect gift. From a visual standpoint, it is a beautiful book, a pleasure to handle and to study. The photographs are both rare and fascinating. As a biography of the pioneer French animator, the book could be no better. The text is serious yet entertaining and all facts are scrupulously annotated. The author notes that Emile Cohl was born in Paris and was "an active participant in its popular culture and was in turn an influence on his milieu." Above and beyond documenting Cohl's work and his influence on French culture, the book discusses the animator's work as it is relevant to the American film industry, the latter's impact on Cohl and Cohl's inspiration for its animators.

I was particularly pleased to find discussion of the work of British pioneer, Arthur Melbourne-Cooper, who has received such short shrift among film historians in his own country. I do not necessarily agree that British historian and theoretician Barry Salt is the best source for information on Melbourne-Cooper, but at least Salt was willing to acknowledge the man's existence and author Donald Crafton was willing to search out relevant facts.

Donald Crafton has already given us a definitive study of early animation, *Before Mickey: The Animated Films 1898-1928.* With his new book, he proves himself to be one of the most preeminent of American film scholars. Highly recommended.

Blacks in Film and Television: A Pan-African Bibliography of Films, Filmmakers, and Performers compiled by John Gray. $55. (Greenwood Press. Hardcover. 512 pages. Index.)

This important bibliographic source book indexes articles and books relating to Blacks throughout the world by countries, individuals, and films. General works are also listed. An extraordinary amount of work has gone into the preparation of this volume, and I do not wish to denigrate the compiler's achievement. However, there are omissions, for example Helen Arstein and Carlton Moss's 1950 biography of Lena Horne is not included. But, what is here will prove to be of lasting value to any researcher or scholar studying Black art and culture as they pertain to film and television.

The Screening of America by Tom O'Brien. $21.95. (Continuum. Hardcover. 219 pages. Index. Illus.)

Tom O'Brien, a former film critic for the Catholic journal *Commonweal,* examines the ethical and moral values to be found in films of the past two decades. He tries to explain why some professions are depicted as glamorous and why others are not, how films have treated women and minorities, why environmental issues are subjects of concern, and why religious leaders are subjects of satire. O'Brien's style is easy to read, perhaps a little too easy. You might agree with an opening comment, "Was there ever a better baseball movie than *Bull Durham*"—I do—but does such a sentence really make you want to read more? Probably not, and that is a pity because this is a book which I like, and it is very obvious that it is based on a first-hand examination of the films themselves.

Universal Horrors: The Studio's Classic Films, 1931-1946 by Michael Brunas, John Brunas, and Tom Weaver. $45. (McFarland. Hardcover. 616 pages. Index. Illus.)

Following a brief but informative history of Universal, the authors offer lengthy essays and detailed credits for 85 Universal horror features, produced between 1931 and 1946. The first film discussed is *Dracula* and the last is *The Brute Man.* Appendices cover Universal's sound serials, a listing of Universal television packages of horror features, and the few Academy Awards won by the features under discussion. The authors provide an extraordinary amount of information on each film, based to a large extent on production files at the University of Southern California (USC) and, apparently, interviews with some of the personalities. I might have wished that the authors had indicated their sources, because sometimes the known source can make all the difference between accepting a fact at face value or disputing its accuracy. The writing style is highly entertaining and sometimes amusing, such as referring to *Murders in the Rue Morgue* as director Robert Florey's "finest hour (and two minutes)." Recommended.

The Films of G. W. Pabst: An Extraterritorial Cinema, edited by Eric Rentschler. $17.95/$45. (Rutgers University Press. Paperback/Hardcover. Index. Illus.)

This, the second book in English on the talented and sometimes discredited German director, G. W. Pabst, is a series of essays covering almost all of his films, together with a discussion by the editor on Pabst as an auteur and a thorough biographical essay by Hans-Michael Bock. The book also includes a very detailed filmography and an even more detailed (if possible) bibliography for each film. Unfortunately, much of the writing is extremely dull and difficult for anyone but the most devoted academic to read. Most of the text reads as if it was translated from the German, and, in fact, it was. Even when an essay is not translated from the German, it still reads as if it was! Recommended with reservations.

Television Weathercasting: A History by Robert Henson. $35. (McFarland. Hardcover. 200 pages. Index. Illus.)

I suppose if there can be a cable channel devoted to the weather, there can also be a book-length study of weatherman (and women), complete with photographs, in radio and television. Mr. Henson takes the subject very seriously, and does not really approve of the Willard Scott approach to weatherforecasting. Mr. Henson does take note that "To his credit, Scott is meticulous about mentioning areas with threatening or extreme weather, even to the point of noting specific severe-weather watches in effect." Bully for Willard Scott, although if I was in an area of severe-weather watching, I doubt I would be basing my reactions on the latter's prognosis. There is an extraordinary amount of information contained in this book, including biographies of weatherpersons and a listing of holders of the American Meteorological Society Seal of Approval for Television. Did you know they actually give out Emmys for excellence in local weathercasting? I don't know who would want to read this book, apart from weatherforecasters, but I am going to award it a Recommended label, if for no other reason than it is the most unusual television-related book of the year.

Berlin, Kern, Rodgers, Hart, and Hammerstein: A Complete Song Catalogue by Steven Suskin. $55. (McFarland. Hardcover. 336 pages. Index.)

For each of five composers and/or lyricists, Irving Berlin, Jerome Kern, Richard Rodgers, Lorenz Hart, and Oscar Hammerstein, the author has compiled brief biographies followed by song listings, production summaries and collaborator summaries. The information on each song includes the date of copyright, date of publication if different, and the production in which the song was first used. The work involved in compiling this song catalogue was tremendous, and I will not even attempt to explain some of the problems faced by Steven Suskin and the reason why it was impossible to make the present volume totally complete. In fact, if the author had not indicated there were omissions, I doubt anyone would have known. Apart from anything else, it is fascinating just to read through the song titles and experience the number of compositions. 930 songs are listed for Irving Berlin. Hammerstein wrote the lyrics for 636 songs. An extraordinary labor of love by Steven Suskin, and, of course, recommended.

And thank you, Mr. Suskin, for spelling "catalogue" the correct way. There is still hope for the English language!

Deadly Illusions: Jean Harlow and the Murder of Paul Bern by Samuel Marx and Joyce Vanderveen. $19.95. (Random House. Hardcover. 271 pages. Index. Illus.)

Written in the first person of Samuel Marx, *Deadly Illusions* provides a journey through the career of Marx as a screenwriter, producer, and film book author, together with his investigation into the death of Paul Bern which he proves (with reasoned argument) to be murder. The murderer committed suicide shortly after killing Bern. Along the way, Marx introduces his collaborator, an actress and ballerina of whom I had not heard before. Marx is a little unkind about Harold Grieve, whose only fault is that he is old and forgetful. Marx is also somewhat outspoken with regard to Grieve's wife, Jetta Goudal, whom he claims had a highly publicized love affair with Paul Bern. This is news to me. I have never seen any contemporary publicity about this. Indeed, having known Jetta Goudal for probably fifteen years, I cannot imagine her having a love affair with anyone except herself—she would be too concerned about messing up her hair!

What I found disturbing about this book is Marx's total lack of concern for his own involvement in the Paul Bern affair. He knew at the time that M-G-M, led by Louis B. Mayer, was discrediting Bern's memory by claiming he was impotent. He did nothing about it then and has waited over fifty years before deciding it is time for the truth.

American Film Music: Major Composers, Techniques, Trends, 1915-1989 by William Darby and Jack Du Bois. $55. (McFarland. Hardcover. 623 pages. Index. Illus.)

American Film Music offers detailed chapters on American film music from the silent era through the 1980s, together with chapters on a group of individual composers: Max Steiner, Alfred Newman, Franz Waxman, Erich Wolfgang Korngold, Hudo Friedhofer, Dimitri Tiomkin, Victor Young, Miklos Rozsa, Bernard Herrmann, Alex North, Elmer Bernstein, Henry Mancini, Jerry Goldsmith, and John Williams. Title listings of films follow the chapters on individual composers, and the book also includes a listing of Academy Award winners for Best Score and Best Song and bibliographical notes on each composer. A major work of reference, *American Film Music* even reproduces relevant bars of music and in so doing is perhaps a little too technical for many readers. No one interested in film music can afford to be without this book. Recommended.

The Merchant of Dennis the Menace by Hank Ketcham. $24.95. (Abbeville Press. Hardcover. 276 pages. Index. Illus.)

Dennis the Menace dates back to 1951, when children could be both mischievous and fun. Today, most kids are not even mischievous—they are downright evil. Creator Hank Ketcham has put together a sumptuous autobiography, replete with many Dennis the Menace cartoons and a goodly number of additional cartoons and photographs. In a text which is somewhat conservative and unimaginative, Ketcham discusses his life, the creation of the Dennis the Menace character—his wife once called their son Dennis a menace—and how he goes about creating his cartoons. The television incarnations (live and animated) are touched upon, and Ketcham pays considerable attention to his life in Hollywood, from 1939 through 1942, when he worked for both Walter Lantz and Walt Disney. Ketcham spent a few days on the picket line during the Disney strike and then returned to work, branded by his fellow workers, King of the Finks. Interestingly, he writes that the strike put Disney "into a seven-year snit." It didn't do much either for those poor souls in the union.

The Merchant of Dennis the Menace is an elegant book. Even with 250 illustrations, it is far, far more than a mere picture book. It's a Christmas gift which will make all those people who don't agree with the philosophy of the Doonesbury cartoons happy.

Errol Flynn: The Spy Who Never Was by Tony Thomas. $18.95. (Citadel Press. Hardcover. 186 pages. Index. Illus.)

Tony Thomas is one of the leading critics of Charles Higham's biography, *Errol Flynn: The Untold Story.* Like most Flynn fans, Thomas is outraged by Higham's depiction of the film star as a Nazi. Unlike most of his colleagues, Thomas sets out to prove Higham wrong, and he has done a splendid job of branding *Errol Flynn: The Untold Story* as highly questionable in parts and flawed with errors. Thomas is very civilized in the manner in which he attacks Higham; he writes of Higham's life and career with respect and humor, calling him "a well educated Englishman" and "clearly a gifted writer." At the same time, Thomas examines the same documents studied by Higham and discovers that the latter has sometimes changed wordings or misinterpreted meanings. He talks to Flynn's friends and colleagues, including Olivia de Havilland, David Niven, Eddie Albert, and Patric Knowles, who contributes an "Epilogue—Rebuttal for a Friend" to the book. He even corresponds with Flynn's Nazi friend Hermann Erben. Thomas publishes a description of a last meeting between Olivia de Havilland and Errol Flynn, followed by Higham's account of the same meeting, followed by de Havilland's rebuttal of Higham's description!

Errol Flynn: The Spy Who Never Was will please all Flynn's admirers. Certainly, it is difficult to dispute the evidence which it presents. In addition, the book offers an entertaining biography of the actor, which does not hide any of his faults, but also gives one the impression of a man who could still be likeable and was always a good friend.

The Great Combat Pictures: Twentieth-Century Warfare on the Screen by James Robert Parish. $47.50. (Scarecrow Press. Hardcover. 486 pages. Illus.)

James Robert Parish has performed his usual reliable job in documenting 300 American and British feature films and made-for-television movies showing soldiers in action from the 'teens through the present. For each production, Parish provides complete credit information and a complementary essay discussing the plot, contemporary reviews, and the film's importance. The amount of space devoted to each film is indicative of its importance—*Aces High* receives one page, while *Full Metal Jacket* has three. The book concludes with a listing of the films by war, and a listing, compiled by Vincent Terrace, of Combat Programs on Television. A useful reference work, and I was very pleased to note that it is dedicated to the memory of that fine screenwriter and chronicler of Hollywood history, DeWitt Bodeen.

The Great Detective Pictures by James Robert Parish and Michael R. Pitts. $59.50. (Scarecrow Press. Hardcover. 630 pages. Illus.)

Similar in format to the previous volume, *The Great Detective Pictures* covers 350 feature films of the genre, from *Accomplice* (1946) to *Young Sherlock Holmes* (1985). The inclusion of the latter indicates the authors are using the word "Great" in a very general sense—personally, I would only have included *Young Sherlock Holmes* in a book titled *The Worst Detective Pictures.* The book concludes with a listing of radio and television detective programs, compiled by Vincent Terrace, a bibliography of detective fiction sources by T. Allan Taylor, and a chronology for the films included in the book. Yet another useful reference work from the Parish and Pitts research and writing workshop.

Movie Anecdotes by Peter Hay. $19.95. (Oxford University Press. Hardcover. 312 pages. Index. Illus.)

Movie Anecdotes is exactly what its title suggests, a collection of one-liners and longer, highly quotable comments on Hollywood and the film industry. Typical of the material here is the following:

An Australian writer visiting Universal Studios was introduced to Carl Laemmle, who asked how long he had been in America.

"Two weeks," said the visitor.

"It's amazing," said the mogul, "how well you speak English after only two weeks."

None of the sources of the material are indicated, and so this is not, in any sense, an academic book—in fact, it is not the sort of book one would expect from the Oxford University Press. The style carries over from the author's two earlier volumes, *Theatrical Anecdotes* and *Broadway Anecdotes.* I suspect this will prove a pleasing Christmas gift for many film buffs.

The Complete Actors' Television Credits, 1948-1988, 2nd Edition, Volume 2: Actresses by James Robert Parish and Vincent Terrace. $49.50. (Scarecrow Press. Hardcover. 447 pages. Index. Illus.)

A companion piece to the earlier volume covering the television work of actors, this book documents television appearances by 1,739 actresses (or so I am told, because I did not bother to count). In chronological order the work of each actress is listed in both television series and made-for-television movies. Following the show title, the authors have provided the date(s) it was aired and the network. Work on PBS and cable television is also included. This is a superlative compilation, an amazing work of reference. Recommended.

The Authorized Guide to Dick Tracy Collectibles by William Crouch and Lawrence Doucet. $12.95. (Wallace-Homestead. Paperback. 162 pages. Illus.)

A little late perhaps for the Walt Disney release of *Dick Tracy* starring Warren Beatty and Madonna, but still more than welcome is this fascinating survey of Dick Tracy memorabilia. Everything is documented here, much of it with photographs. There are books, comic books and artwork, guns (!), games, badges, recordings, and, of course, two-way wrist radios. All of the items in the main body of the book are numbered, and a corresponding price guide is provided in the back of the volume. The book is beautifully produced—a credit to its publisher and its authors—and includes an introduction by the daughter of Dick Tracy's creator, Jean Gould O'Connell.

Hollywood Gothic: The Tangled Web of Dracula from Novel to Stage by David J. Skal. $39.95. (W. W. Norton. Hardcover. 242 pages. Index. Illus.)

Not another book on horror films? No! Not just another book, but the best of its type, a superb study of the various incarnations of Bram Stoker's *Dracula* on the printed page, on stage, and on screen. The author's research is impeccable, involving the study of studio records, interviews with contemporary figures (notably Raymond Huntley and Lupita Tovar), examination of legal papers (the law suit between Bela Lugosi, Jr., and Universal is very well documented), and even a U. S. Treasury Department sanctioned trip to Cuba to view the only surviving, complete print of the Spanish-language version of the 1930 *Dracula.* (The author devotes considerable space to this film and is obviously very impressed with it.)

Hollywood Gothic is a pleasure to hold and to read, beautifully designed and including many illustrations which I have never before seen (including frame enlargements from the aforementioned *Dracula*). A useful chronology complements the volume. I might also add that despite its easy reading style, *Hollywood Gothic* can also be considered a work of academic scholarship, in that all statements are carefully footnoted. This is the perfect Christmas gift for anyone interested in the horror genre. It is well worth seeking out at bookstores, but be warned, Crown Books is refusing to stock it. Highly recommended.

Webster's New World Dictionary of Media and Communications by Richard Weiner. $29.95. (Webster's/Prentice Hall. Hardcover. 533 pages.)

Perhaps as evidence of the coming of age of the communications industry (broadcasting, film, and the print media), Webster's has published a dictionary providing brief definitions or explanations of terms as varied as Contouring, Limpet Mount, Pan and Scan, Return Key, and Alan Smithee. The last, in case any reader is unfamiliar with the name, is used in film or television credits when a director does not want to be identified, usually because he or she has had an argument with the producer. Some of the entries are, conceivably, a little short on substance—only 50 words on the Academy of Motion Picture Arts and Sciences—but what the dictionary lacks in depth, it makes up in preciseness and brevity. In fact, over 30,000 terms are defined here in a book which is obviously going to become a standard work of reference. Incidentally, the author introduced the Cabbage Patch Kids! Recommended.

The Casting Couch and Other Front Row Seats: Women in Films of the 1970s and 1980s by Marsha McCreadie. $38.95. (Praeger. Hardcover. 208 pages. Index.)

Marsha McCreadie has gathered together a collection of her writings from the last two decades, primarily from *The Arizona Republic,* which are, surprisingly, highly readable. In fact, the most remarkable aspect of this anthology is that a publisher would be willing to produce a book on a feminist topic which a reader can enjoy and which does not require a dictionary at one's right hand while trying to wade through it. Ms. McCreadie has grouped her essays/reviews under topic headings such as Female Bonding and New Genres. She has also included a collection of short profiles on women filmmakers from Lillian Gish to Gilda Radner, some of which could do with a little more depth in the analysis or the subject's comments. However, on the whole this is a pleasing collection of items and a useful overview of films from the Seventies and Eighties dealing with women's issues.

1991

January 1991

Encyclopedia of American Spy Films by Larry Langman and David Ebner. $49.95. (Garland. Hardcover. 443 pages. Index. Illus.)

Encyclopedia of American Spy Films is the type of book you expect to see James Robert Parish listed as author of—an awful lot of books these days have Parish's name on them, but this is not one of them. One of the authors, Larry Langman, co-authored *Encyclopedia of American War Films,* and the current offering is very similar in format. It includes essay-entries on hundreds of films, on each of which the authors provide basic credits, synopsis, and an occasional critical comment. Also included are entries on various topics associated with the subject, and these include "Anarchists in the U. S.," Mata Hari, Pearl Harbor, War of 1812, and dozens of others. In fact, it is these entries which make the book both unusual and more valuable. Appendices list all of these "topic" entries, as well as spy serials, spy films which have received awards or nominations from the Academy of Motion Picture Arts and Sciences, the New York Film Critics, and the National Board of Review, spy films about political assassinations, and spy films listed by their related wars. Some of the entries are disappointingly brief, but nevertheless this is a useful reference work.

Babe: The Life of Oliver Hardy by John McCabe. $16.95. (Citadel Press. Hardcover. 224 pages. Index. Illus.)

Fifteen years after publication of his definitive biography of Stan Laurel, John McCabe has produced a similar tribute to Mr. Laurel's partner. One expects impeccable research, authoritative opinions, and an elegant writing style from Mr. McCabe—and he does not disappoint. Also, although the credit should not go to the author, how nice to find a hardcover book which is reasonably priced for a change—this biography is cheaper than many paperbacks of similar length! All aspects of the comedian's life and work are documented here, including the years in which Oliver Hardy was not partnered by Stan Laurel. Perhaps from a morbid curiosity, I always find the later (and arguably lesser) years of a personality's life more interesting, and the author has not scrimped in telling of the couple's music hall performances in England and Ireland in the late 1940s and early 1950s, as well as the making of *Atoll K.* (From a purely personal viewpoint, I was intrigued to learn that McCabe first met Laurel and Hardy in my hometown, Birmingham, England, and that his second meeting with the pair was in Aston, a suburb of Birmingham, where I grew up.) The book also includes a basic filmography, listing only film title, production company, complete date of release, and number of reel, from *Outwitting Dad,* produced by Lubin in 1914, through *Atoll K,* released in the United States in 1954 as *Utopia.*

I suppose if you want to find cause for complaint, it is that John McCabe is too loving in the handling of his subject's life. But, at the same time, it is very obvious from reading this biography that Oliver Hardy gave only love and with love should he be rewarded. Highly recommended.

Collector's Guide to Autographs by George Sanders, Helen Sanders, and Ralph Roberts. $16.95. (Wallace-Homestead. Paperback. 225 pages. Index. Illus.)

If anything, this book proves that autograph collecting is not a subject that warrants a book-length study. With the exception of a piece by Helen Sanders on the discovery of a collection of Garbo items, and an interesting discussion on the use of the autopen by celebrities, most of the book is devoted to the obvious. While the whole spectrum of autograph collecting is covered, movie-related autographs receive prominence. In part, this is because *Collector's Guide to Autographs* contains innumerable photographs of one of the authors, George

Sanders, interviewing personalities primarily from the entertainment field. Sanders (no, he's not the well-known actor) was, apparently, involved in radio and television, and also an actor in more than 20 films. (As an aside, I thought the Screen Actors Guild did not permit two actors with the same name.) Obviously, I am not familiar with Sanders the actor; I do know him as co-owner of the Autograph House, a popular dealer in autographs. I am happy to recommend his company, but as far as the book is concerned, it is obviously for the newcomer or beginner.

Hake's Guide to TV Collectibles by Ted Hake. $14.95. (Wallace-Homestead. Paperback. 185 pages. Index. Illus.)

Ted Hake is a well-known dealer in all types of late 19th and early 20th Century Americana. A one-man popular culture program, Hake's catalogs always make fascinating reading. In this book, he provides a first-rate, highly detailed survey of television-related memorabilia. The book is arranged alphabetically by television series. Under the title of each series, Hake indicates the network and the original dates when the program was broadcast. Then, he lists and describes the memorabilia generated by the show, together with its current (1989) value. *Howdy Doody* is the earliest program listed (dating back to 1947), and it has also generated the largest amount of memorabilia—31 items. Far more typical is *The Adventures of Robin Hood,* starring Richard Greene. Its success resulted in a metal lunch box, a jigsaw puzzle, a game, and "Robin Hood Shields." Nothing could be more sensible than the manner in which this book is arranged. It is easy to use, the descriptions are succinct, and the prices (because of the author's reputation) are obviously reliable. The book also contains two color sections, and a list of television fan clubs, with addresses. A superb piece of work, and recommended.

American Animated Films: The Silent Era, 1897-1929 by Denis Gifford. $29.95. (McFarland. Hardcover. 160 pages. Index.)

A companion work to the same author's 1987 book, *British Animated Films, 1895-1985,* the present volume lists all of the animated films of the American silent era by series. Under each series heading, Denis Gifford indicates distributor, production company, producer, director, and animator (where known), and then lists the films within the series with release dates (day, month and year). All of the titles in the series are indexed and easily accessed. Quite obviously, this is a tremendous labor of love, and one should be very grateful to its compiler for his dedication. If the book has a fault it is that Gifford fails to indicate his sources. In that information on some of these early animated shorts is very sketchy, it would have been helpful to know how accurate are these listings, and the only way that a reader can establish that is by knowing the primary materials used. At least a bibliography could have been included. Recommended.

Metaphor and Film by Trevor Whittock. $39.50. (Cambridge University Press. Hardcover. 178 pages. Index.)

Metaphor and Film is one of those books! You know what I mean? Its author writes about the suggestive meanings to be found in feature films, consciously introduced by their directors. You know what I mean? It discusses syntagmatic and paradigmatic. You know what I mean? It has pages and pages devoted to the eidophor, the diaphor, synedoche, objective correlative, and chiming. You know what I mean? What, you don't know what I mean! You don't know what the author means! Well, neither do I! And don't bother to get out your *Webster's Dictionary*, because these words aren't in there! But, according to the publisher, if you buy this book, the author will provide an analysis of the mental operations that one must bring to an understanding of the subject. Then, of course, you can just forget about *Metaphor and Film* and be an ignoramus like Samuel Goldwyn, who once said that messages are for Western Union.

Fashion in Film edited by Regine and Peter W. Engelmeier. $49.95. (te Neues. Hardcover. 245 pages. Index. Illus.)

There can be no doubt that *Fashion in Film* is a beautiful book. Printed on fine quality art paper are almost 400 "glamor" photographs of male and female screen personalities from the 1920s through the present. A number beside each photograph refers the reader to the back of the book, where the editors have provided very brief information as to the costume designer, etc. Many times, the photographs are head-and-shoulder shots, which are totally irrelevant in terms of costume design, and so this is very much a pretty book of pretty pictures with little connection to the often harsh and far-from-glamorous world of costume design. In fact, it is really a book about portrait photography. On the plus side, *Fashion in Design* does contain a three-page essay by Audrey Hepburn on "The Costumes Make the Actors" and one-paragraph biographical sketches on some 14 costume designers. All the other essays are worthless, and, yet again, irrelevant to the subject of the book.

Movie Buff Checklist by Marvin Jones. $19.95. (Campfire Video. Paperback. 205 pages. Index. Illus.)

Movie Buff Checklist is, without question, the most extraordinary reference work to arrive on my desk for review. It is an extensive listing of nude appearances by male actors on screen from the 1920s to the present. A brief introduction surveys the field, and then the author lists each actor with the title of the film, the date it was released, the nature of the nudity (rear, frontal, side), and the circumstances under which the nudity occurred. Unknown and groups of players are also included in the listing. Profusely illustrated, *Movie Buff Checklist* makes fascinating reading. For example, it reveals that Malcolm McDowall has made more nude appearances than any other actor—eight, half of which were frontal. If your desire is a frontal nude view of Gregory Harrison, you will have to rent the 1973 feature, *The Harrad Experiment,* in which Harrison was an often-nude extra. Don Johnson was the frequently nude star of the film, and both he and Harrison are exposed in photographs in this book. Other actors caught in the frontal nude state in photographs are Sylvester Stallone, Jan-Michael Vincent, Peter Firth, Sam Jones, Alan Bates, and Johnny Crawford. Bruce Davidson, who has been seen nude from the rear three times on screen, provides an amusingly honest foreword.

This is one book that a lot of film buffs are going to want copies of, if not entirely for the right reasons.One quote on the book's jacket suggests, "Leonard Maltin and the *Movie Buff Checklist* are the two indispensable books beside the VCR" Hm! For no other reason than its uniqueness, I am going to award *Movie Buff Checklist* Recommended status.

The London Stage 1930-1939: A Calendar of Plays and Players by J. P. Wearing. 3 vols. $137.50. (Scarecrow Press. Hardcover. 1,999 pages. Index.)

These three volumes represent the fifth in a series of calendars providing a chronological listing of all plays on the London stage from 1890 to the present. The current volumes document 4,256 productions seen at 61 West End theatres between 1930 and 1939. Full cast and technical credits are provided for each production, along with the dates the play opened and closed, the theatre, the number of performances, and the dates of reviews in various publications. If nothing else, these books prove that London was, is, and always will be the theatre capital of the world. They also represent a mind-boggling amount of research by their author, for whom I have nothing but praise and admiration. Despite the hefty price, no one interested in theatre history can be without these volumes. Recommended.

On Location in Lone Pine by Dave Holland. (The Holland House. Paperback. 106 pages. Index. Illus.)

One early Hollywood producer came up with the statement, "A rock is a rock, film it in Griffith Park." Well, this new book proves that one rock is not always the same as another, as the author identifies the various locations around Lone Pine, California, where such film as *Gunga Din, Riders of the Purple Sage, Lives of a Bengal Lancer, How the West Was Won, Nevada Smith, Romance of the Rio Grande,* and countless others were filmed. The author provides maps for locating the various movie sites, photographs of the sites as they appear in films, an extensive text, and a complete listing of films and television shows shot in whole or part in the area. I am not into rocks, but Dave Holland certainly knows his. He details how he was able to find these movie locations, and explains how the reader can identify a location site by the rock formations, which have not changed through the years.

On Location in Lone Pine is the perfect companion for anyone driving through this beautiful and rugged area of California. It is also a "must" for film buffs and reference librarians. Happily, it joins a growing number of similar books written by devoted enthusiasts determined to identify movie locations. Finally, as the author notes, "That's what this book is all about: taking you to some of those magical places, standing there together and holding hands with the past." Highly Recommended.

John Wayne: Actor, Artist, Hero by Richard D. McGhee. $35. (McFarland. Hardcover. 400 pages. Index. Illus.)

As its title suggests, *John Wayne: Actor, Artist, Hero* considers the career of John Wayne in a variety of categories. The author looks at Wayne as a tragic hero, a comedian, a romantic hero, a patriot, a multifaceted character, and much more. The research is commendable and the writing style attractive. There can be no doubt that Richard D. McGhee knows his subject's films well and is able to channel his appreciation of them to the reader. The book also includes a first-rate filmography.

Curiously, the publisher considers it necessary to identify the author as "a member of the Cherokee Nation of Oklahoma." I do not understand the relevance of this. As a reviewer, I do not care to know the author's ethnic background, the color of his skin, his religion, his nationality, or his sexual orientation. I consider all authors as members of the human race trying to express their thoughts on paper, succeeding or failing to interest the reader in what they have to offer. Nothing else matters. Having said that, I am happy to recommend *John Wayne: Actor, Artist, Hero,* not because the author is a member of the Cherokee Nation of Oklahoma, but because he has written a first-rate book.

Early Black American Playwrights and Dramatic Writers: A Biographical Directory and Catalog of Plays, Films, and Broadcasting Scripts by Bernard L. Peterson, Jr. $65. (Greenwood Press. Hardcover. 328 pages. Index.)

The main body of this book consists of biographical entries on 136 Black American playwrights. For each, the author provides a brief biography, together with listings of works for stage, screen, and occasionally the broadcast media, together with additional sources for further reference. While primarily devoted to Black Americans who received recognition for their stage work, a number of prominent names from the film world are included, notably Carlton Moss and Clarence Muse. Following the main section, the author includes an appendix on "Other Early Black American Playwrights and Their Plays," entries on additional musical librettists, a chronology of Black-authored works in the performing arts from 1823 to 1950, and listings of information sources with a lengthy bibliography.

The book is a companion volume to the author's 1988 work, *Contemporary Black American Playwrights and Their Plays: A Biographical Directory and Dramatic Index,* and to an upcoming volume to be devoted to the musical stage. An important reference source, which is doubly helpful in that immediately it directs the reader to additional sources of information.

Television Network Daytime and Late-Night Programming, 1959-1989 by Mitchell E. Shapiro. $39.95. (McFarland. Hardcover. 264 pages. Index.)

A companion volume to the author's *Television Network Prime-Time Programming, 1948-1988* (published in 1989), the current work covers daytime and late-night programming on ABC, CBS, NBC, and Fox from September 1959 through the fall of 1989 and is another valuable reference source for television research.

The Great Cop Pictures by James Robert Parish. $62.50. (Scarecrow Press. Hardcover. 693 pages. Illus.)

I opened this month with an ersatz James Robert Parish book. I close with the real thing. Parish's latest offering is devoted to feature films and made-for-television movies dealing with the American police forces, or "cops" (a regrettable slang expression which the author and publisher use). Some 330 films are covered from the 1920s through the present, and for each, Parish has provided detailed credits and a commentary cum synopsis. He also quotes from reviews and sample dialogue. I don't think the latter serves much purpose, as, certainly in the case of *Police Academy 4,* it hardly justifies the author's opinion that the film has "a few bright satirical moments." The book also includes a listing of "Cop Series" on radio and television, compiled by Vincent Terrace, a chronology, and a selective bibliography of books on police by T. Allan Taylor. As with all of Parish's volumes, the research is reliable, and the book serves as a useful and speedy resource for basic material on the films covered.

February 1991

Behind the Mask of Innocence by Kevin Brownlow. $50. (Alfred A. Knopf. 579 pages. Index. Illus.)

Back in 1968, Kevin Brownlow published his first volume on the American silent film, *The Parade's Gone By.* It was followed in 1979 by *The War, the West and the Wilderness.* Now, Brownlow has completed his trilogy on American silent films with *Behind the Mask of Innocence.* All three books represent some of the most important writings on the subject. In fact, I would go out on a limb and suggest that the last two volumes, which have not, and probably will not, garner the same praise as *The Parade's Gone By,* are the more important, and, arguably, unique. The research here, as before, is impeccable, the concern for detail obvious, and the author has such a pleasant writing style that his obvious love of, and concern for, silent films is infectious, even for the non-enthusiast.

Behind the Mask of Innocence takes its title from Edward Wagenknecht's 1962 book, *The Movies in the Age of Innocence.* Here, Kevin Brownlow provides a detailed survey of the films which Wagenknecht probably would have seen, or been moved by, as a child—the films of social consciousness. In a heavily annotated survey, Brownlow discusses early attempts at film censorship and then surveys American films dealing with such varied societal issues as birth control, white slavery, drug addiction, prohibition, crime, political corruption, women's suffrage, prison conditions, poverty, ethnicity, anti-Semitism, child labor, and capital versus labor. In each area, he examines specific films, detailing the plots, contemporary reactions, occasionally quoting from personalities involved (but not as much as in *The Parade's Gone By*) and offering his own views.

From a personal viewpoint, I was pleased to read Brownlow's words of praise for Lois Weber—the "only filmmaker in America [who] devoted an entire career to making what were known as 'thought films' "—and his comments on the outrageously amusing Mary MacLane and her film, *Men Who Have Made Love to Me.* Mary

MacLane, along with many of the white slavery features, represents the lowest points in American social problem films, but, of course, much of the book is concerned with more serious areas of study and more thoughtful films, such as the screen version of Upton Sinclair's *The Jungle* (there is a beautiful color poster from this film reproduced on the book's dust jacket), the Jewish themes in the works of director Edward Sloman, and Mrs. Wallace Reid's *Human Wreckage*.

Unlike many authors of film books, Kevin Brownlow clearly has a knowledge of what was happening in the United States at the same time that the films which he discusses were being made. He links the production of many of the films to national and, sometimes, international movements.

This is a hefty tome and is, perhaps, priced a little high for many film buffs. Nevertheless, it is a book which no one should be without. Highly Recommended.

1939—The Year in Movies: A Comprehensive Filmography by Thomas D. Flannery. $39.95. (McFarland. Hardcover. 302 pages. Index. Illus.)

Back in 1989, Ted Sennett produced a sumptuous picture book devoted to the films of 1939. Now, two years later, McFarland has published a more scholarly examination of 125 feature films of the year. Both American and foreign films are included, but the author gives no indication as to how he selected the latter. The more important films receive detailed credits, a synopses, a commentary, and selections from contemporary reviews. Minor films receive a lot less. The listing for *Nurse Edith Cavell* includes only four lines of text and a listing of the principal players, production company, running time, and director. Gene Autry's *Mountain Rhythm* receives the same credit information, but only two lines of text. The author provides a complete listing of the year's Academy Award nominees and winners, together with a breakdown of what took place at the ceremony. He also offers a loving, two-page tribute to his hometown movie theatre, the Irving in Carbondale, Pennsylvania. A production error has resulted in the captions for the still from *The Little Princess* being transposed—at least I hope that is what happened and not that the author cannot differentiate between Richard Greene and Arthur Treacher!

1939—The Year in Movies is not a great book, but it is a useful reference source. Certainly, it further demonstrates just how important a year in film history this was.

Trevor Howard: The Man and His Films by Michael Munn. $19.95. (Scarborough House. 194 pages. Index. Illus.)

"I'd hate to think I'd only ever be remembered for 'Brief Encounter.' Or for being a hellraiser; or for working with Marlon Brando," Trevor Howard once remarked. This new book, first published in England and replete with such English phrases as "chip butty," attempts to provide a more rounded picture of the actor, who was a heavy drinker if not a hellraiser, called everyone "amigo," loved cricket, and hated working on *Mutiny on the Bounty, Ryan's Daughter* and quite a few other films, including *Pope Joan.* "God knows why I got involved," Howard is quoted, "Must have forgot to read the script!"

Unfortunately, *Trevor Howard: The Man and His Films* is a pedestrian, rather hurried biography of the star, full of information obviously gleaned from publicity releases, and occasional quotes from co-stars. It lacks the detailed research one has come to expect in many American biographies, such as those by Charles Higham, and is written in a style which tends to bore rather than attract the reader.

Small-Town America in Film: The Decline and Fall of Community by Emanuel Levy. $27.95. (A Frederick Ungar Book/Continuum. Hardcover. 298 pages. Index. Illus.)

Decade by decade, Emanuel Levy discusses the depiction of small-town American life on screen, from the 1920s through the present. He ignores the silent era except for comments on two features, King Vidor's *The Crowd,* and F. W. Murnau's *Sunrise* (which he discusses in relationship to Vidor's *Our Daily Bread).* Feature after feature is discussed in considerable detail and Levy has more than a fair amount of interesting comments to make—his study of *Blue Velvet* is first-rate. The author concludes with the rather obvious statement "that, along with Westerns, narratives about small towns have been uniquely *American* films." He also notes the number of different genres that have been utilized by filmmakers in depicting small-town life, and the book may be considered an argument for recognizing "Small-Town America" as a genre in its own right. *Small-Town America in Film* also includes a six-page listing of small-town films, decade by decade.

Audio Post-Production in Video and Film by Tim Amyes. $34.95. (Focal Press. Hardcover. 204 pages. Index. Illus.)

Written from a British perspective—the author is a sound supervisor with Scottish Television—*Audio Post-Production in Video and Film* offers a thorough explanation of the techniques involved in post-production sound. The book includes a useful glossary of terms (providing an answer to the often asked question in film credits as to what is a Foley Editor?), many sketches to supplement the text, and a rather unsatisfactory "historical background" to the subject.

A Youth in Babylon: Confessions of a Trash-Film King by David F. Friedman with Don De Nevi. $19.95. (Prometheus Books. Hardcover. 355 pages. Illus.)

David Friedman presents a surprisingly lengthy and detailed account of his life and career in the exploitation film business up through 1964. *A Youth in Babylon* is immense fun to read, and provides what is surely unique documentation on such legendary fellow showmen as Dwain Esper, Howard "Pappy" Golden, S. S. Millard, and especially Kroger Babb (to whom a considerable portion of the book is devoted). Naturally, Friedman discusses cultist director Herschell Gordon Lewis and the making of their many films together, particularly *Blood Feast.* It says much for Friedman's talents as a storyteller that as the result of reading this book, I actually would welcome the opportunity to view *Scum of the Earth, Two Thousand Maniacs, The Adventures of Lucky Pierre,* or *Mom and Dad.* I'd even be willing to watch the last without the in-person lecture by "Elliot Forbes" on "The Secrets of Sensible Sex."

A Youth in Babylon offers the reader a welcome return to those happy days before *Playboy* and X-rated home video and cable television, when "adults only" entertainment could not bother anyone except the Legion of Decency. The only fault with the book is the lack of an index and Friedman's occasional repeating of stories (probably the result of sloppy editing).

The book closes with the promise of a second volume, *Kings of Babylon,* which will include "for your added pleasure and perusal a plethora of photos—the kind you like to look at—in private!" I can hardly wait!

I Went That-a-Way: The Memoirs of a Western Film Director by Harry L. Fraser, edited by Wheeler W. Dixon and Audrey Brown Fraser. $22.50. (Scarecrow Press. Hardcover. 161 pages. Index. Illus.)

I Went That-a-Way, No. 22 in the Filmmakers Series, is the autobiography of "B" picture director Harry Fraser, responsible for dozens and dozens of Westerns starring Rex Bell, Tom Mix, Lash La Rue, John Wayne, and others, as well as one of my favorite exploitation films, *Chained for Life* with Daisy and Violet Hilton. The book also contains a lengthy introduction by Wheeler W. Dixon and a basic filmography.

Film Flubs: Memorable Movie Mistakes by Bill Givens. $7.95. (Citadel Press. Paperback, 157 pages, Index. Illus.)

Ray Liotta, playing Shoeless Joe Jackson in *Field of Dreams,* bats right-handed—but the real Shoeless Joe was a southpaw. That is one of the hundred or so movie errors documented in this amusing little book. Not a volume that readers will probably want for their reference shelves, but *Film Flubs* is a perfect (and cheap) gift for the average film buff. There are even photographs immediately proving some of the gaffes and anachronisms, which are presented here under various categories, such as "Now You See It, Now You Don't" and "Prop Problems."

The House of Barrymore by Margot Peters. $29.95. (Alfred A. Knopf. Hardcover. 642 pages. Index. Illus.)

Ethel, John, and Lionel Barrymore all published their own autobiographies and have all been the subject of at least one biography. It might, therefore, be reasonable to question the need for another book on the Barrymore clan. Such a query immediately becomes irrelevant when the author is Margot Peters. She has already given us a stylish and elegant biography of Mrs. Patrick Campbell (published in 1984). *The House of Barrymore* is in the same literate tradition, meticulously researched and beautifully written.

James Dean: Behind the Scene by Leith Adams and Keith Burns. $29.95. (Birch Lane Press. Hardcover. 223 pages. Illus.)

At first glance, *James Dean: Behind the Scene* might appear to be nothing more than a picture book on the actor. In reality, it is, arguably, the definitive study of Dean's three features, *East of Eden, Rebel without a Cause,* and *Giant.* Aside from the more than 200 never-before-published photographs, the book contains pages and pages of production data relating to the films, including interoffice memoranda, correspondence, and telegrams. One of the authors, Leith Adams, is archivist of the Warner Bros. Collection at the University of Southern California, and, obviously, he is in a unique position to provide the type of documentation denied or inaccessible to most writers on the subject. I only wish he and his coauthor would have included an index.

In addition, the book includes a four-page introduction by Dennis Hopper, in which the actor states categorically that Dean was not gay and that "my life was confused and disoriented for years by his passing." I suppose that is Hopper's excuse for such lousy films as his 1971 self-indulgent exercise in mediocrity, *The Last Movie.*

James Dean: Behind the Scene more than justifies its authors' description as "our celebration of James Dean."

Cahiers du Cinéma, 1969-1972: The Politics of Representation edited by Nick Browne. $30. (Harvard University Press. Hardcover. 352 pages. Index.)

This current volume gathers together important essays from the legendary French film journal, *Cahiers du Cinéma,* published between

1969 and 1972. Two earlier volumes have covered the years from 1951 through 1968. Obviously the anthology is important in that it translates and makes more widely available the included articles. At the same time, it does require considerable intelligence and concentration from its readers. (And, quite frankly, there are times in reading books such as this that I feel I have neither.) However, for those with time and patience, there are some fascinating pieces here—I was particularly stirred by Jean Louis Comolli's "Technique and Ideology: Camera, Perspective, Depth of Field." All articles are heavily footnoted, and the book includes a solid introduction by its editor, who notes, "What is dramatized in this collection . . . is the spectacular action of rigorous and politically engaged film criticism."

The Book of Video Lists by Tom Wiener. $12.95. (Madison Books. Paperback. 584 pages. Index.)

This is the third edition of a book, first published in 1988, and so it must be popular with videotape users. Personally, I found it rather strange and difficult to understand its value in comparison, say, to Leonard Maltin's *TV Movies and Video Guide.* In fact, two-thirds of this book consists of more than 5,000 entries similar in format to those in Maltin's book, except that Tom Wiener does not rate the films. The front section of *The Book of Video Lists* consists, as the title suggests, of more than 600 lists—of films arranged by varied categories, such as "Modern Problems" and "Zombies," and by stars. (In many respects, and with such a variety of headings, these lists make the book useful as a subject index to popular films.) There are also listings of films that are essential viewing and major films which are not available on video cassette. The author does not provide information as to the distributors of the cassettes, maintaining that this is a book for the videotape renter, relying on his or her video store to provide the desired tape.

George Sanders: An Exhausted Life by Richard Vanderbeets. $21.95. (Madison Books. Hardcover. 271 pages. Index. Illus.)

George Sanders published his autobiography, *Memoirs of a Professional Cad,* in 1960. (It will be reprinted next year by Scarecrow

Press with a new introduction and detailed filmography by Tony Thomas.) Richard Vanderbeets' book is the first biography of the actor, noted for his often contemptuous and cynical attitude towards both his career and life. "Olivier was born with a desire to act," he once commented. "My own desire as a boy was to retire."

George Sanders: An Exhausted Life is well researched and documented, with chapters alternating between Sanders' career and his personal life. Its author has made extensive use of the actor's personal papers and was also assisted by Sanders' sister, Margaret. The actor's second wife, Zsa Zsa Gabor, was also more cooperative than one might have imagined from her recent behavior. She does tend to denigrate, and the author goes along with her, Brian Aherne's friendship with Sanders, and claims, "I don't think Brian knew much about George's private life at all."

The author makes no reference to Sanders' supposed bisexuality, which is asserted by Lawrence J. Quirk in his recent biography of Bette Davis. In fact, according to Vanderbeets, Sanders made a conscious decision to cultivate the image and perpetuate the myth of a woman hater, because it added to his public image as a cad. Such remarks as "all women make better mistresses" and "it would never occur to me to look at a woman before breakfast," are typical of the actor's public comments.

The book concludes with a filmography and a good, solid bibliography. The latter includes Sanders' two mystery novels, published in 1944 and 1946 respectively. Vanderbeets notes that both books were ghostwritten, but is unable to reveal the identities of the authors. I don't think there is any dispute that Leigh Brackett, to whom the book is dedicated, ghosted the second novel, *Stranger at Home,* and the first, *Crime on My Hands,* was ghosted by Georgiana Ann Randolph, who wrote mystery novels under the name Craig Rice. Recommended.

William Friedkin: Films of Aberration, Obsession and Reality by Thomas D. Clagett. $35. (McFarland. Hardcover. 302 pages. Index. Illus.)

On page 279 of this book is a listing of fifty individuals whom Thomas D. Clagett interviewed in compiling his book, which is, I believe, the first to be devoted to the director of such popular features as *The Night They Raided Minksy's, The French Connection,* and *The*

Exorcist. If nothing else, this list of interviewees reinforces my initial opinion on reading this volume—here is the definitive work on its subject. All of Friedkin's twelve feature films and four documentaries are considered in lengthy chapters which not only discuss the making of each film in detail, but also offer a critical evaluation of the work. Certainly, Clagett is perhaps a little too appreciative of his subject, but that has not totally hindered his ability to discuss the films with objectivity. In truth, I cannot share his enthusiasm for Friedkin, whose career seems to have diminished with each new film, but that is my personal opinion and does not alter my high regard for this book.

The volume also includes a bibliography and a very detailed filmography. In that the author is involved in the industry, as an editor, he is able to discuss Friedkin's work in a professional and informed manner. Recommended.

Sunday Nights at Seven: The Jack Benny Story by Jack Benny and His Daughter Joan. $19.95. (Warner Books. Hardcover. 302 pages. Illus.)

This is Jack Benny's autobiography, which his daughter, Joan, unearthed in 1984. In an introduction she explains the reasons why she believes it was not published. Interspersed with the autobiography is commentary by daughter Joan and also included is a foreword by George Burns. The latter describes Benny as "a nice man," and this book strengthens Burns' statement. *Sunday Nights at Seven* is a nice book about a nice man. Nothing more need be said.

Love, Sex, Death & the Meaning of Life: The Films of Woody Allen by Foster Hirsch. $12.95. (Limelight Editions. Paperback. 248 pages. Index. Illus.)

First published in 1981, this is a new edition of Foster Hirsch's study of the life and career (mainly the latter) of Woody Allen. The current edition of the book includes a new final chapter, discussing Allen's films in the 1980s, and an expanded filmography. This is not the only book-length study available on the comedian, but it is certainly the most readable.

If I have a complaint about the book, it is that lumping all of Allen's films of the last decade together makes for a hurried evaluation of them. Almost 200 pages are devoted to Allen's first eleven features, and only 35 to his last eleven, among which were such prominent titles as *Zelig, Hannah and Her Sisters,* and the wonderful *Radio Days.* Perhaps it all goes to prove that Allen's filmmaking career is too complex, too varied to be discussed in one volume.

March 1991

The Power and the Glitter: The Hollywood-Washington Connection by Ronald Brownstein. $24.95. (Pantheon Books. Hardcover. 437 pages. Index. Illus.)

During last year's political campaign in California, actress Angela Lansbury appeared in a television commercial opposing a term limitation initiative. Following a storm of protest, Miss Lansbury withdrew from the commercial, claiming she had been duped—she didn't know what the initiative was about, she didn't understand politics, she never read the newspapers. It was enough to make one television critic suggest a term limitation on actresses in politics.

The subject matter of this new and important book is the relationship between Hollywood and Washington politics, from Louis B. Mayer's support for Herbert Hoover in the early 1930s to liberal Hollywood's unease with Michael Dukakis in the 1980s. It is all here, and fascinating reading: the Warren Beatty-Gary Hart show; the right versus the left as represented by Charlton Heston and Ed Asner; Lew Wasserman and the turnover of power to the newer movie moguls; "the Malibu Mafia" and much more. Jane Fonda receives considerable attention—her father, Henry, refused to let her stay in his home and blamed her politics for damaging his career. John F. Kennedy's impact on the reawakening of Hollywood politics after the McCarthy era is discussed in detail, as is the rise of Ronald Reagan, after George Murphy had proved that Americans were willing to accept actors as politicians.

The Power and the Glitter has a liberal slant, but that does not mean that liberals come off particularly well in the book. What is one to make of Norman Lear, who thought it inappropriate to give anyone but the wealthy the opportunity to question the 1988 Democratic candidates for president? Or Justine Bateman, who asked a Democratic Senator why she had to pay so much in taxes, "Can't we pass a law or something?" Based on the brainless character played by Miss Bateman on *Family Ties,* it is obvious that life does indeed imitate art.

The Power and the Glitter is an important volume, but it would be wrong to suggest it is the definitive study of the subject. Politics and the film community during the silent era are glossed over in a couple of pages. The writer tends to oversimplify matters when discussing film history. For example, he states that in 1927 Warner Bros. married a soundtrack to a moving picture, but, of course, the studio did no such thing—there was no soundtrack, just a Vitaphone disc. However, such faults are minor, and *The Power and the Glitter* will serve our immediate needs both as a textbook history of politics and Hollywood (at least from the 1950s onwards) and as entertaining reading matter. Recommended.

Making Visible the Invisible: An Anthology of Original Essays on Film Acting edited by Carole Zucker. $39.50. (Scarecrow Press. Hardcover. 438 pages. Index. Illus.)

Making Visible the Invisible is a rather tedious collection of original essays on acting for the screen. A variety of topics are discussed, such as "a feminist reading of role playing in Bergman's *Persona* and Mai Zetterling's *The Girls,*" "the confluence of the written word and its articulation in a variety of performance styles in Renoir's *La Grande Illusion,*" and "the repression of projection and identification in the work of Bresson." You get the idea. Most of the essays are the type of piece written by one academic for another—and the potential readership for a book such as this is the number of its contributors. The only essays remotely worth reading are Roberta E. Pearson on "A Semiotic Approach to Acting in the Griffith Biographies" (not as terrible as it sounds), William Rothman's "Virtue and Villainy in the Face of the Camera," and Marian Kean's piece on John Barrymore, "The Great Profile: How Do We Know the Actor from the Acting?" Incidentally, I wish Ms. Keane would explain how a film can be "so deteriorated that it cannot be seen."

Film Review, 1990-1 edited by F. Maurice Speed and James Cameron-Wilson. (Virgin Books. 192 pages. Index. Illus.)

The new (and 46th edition) of the British annual, *Film Review*, is now available from specialized bookstores. Joining F. Maurice Speed as coeditor is James Cameron-Wilson, but the contents remain much the same: releases of the year in the United Kingdom with capsule commentaries and photographs; Anthony Slide's Letter from Hollywood; Movie Quotations of the Year; TV Feature Films of the Year; Video Releases; the Ten Most Promising Faces of 1990; Film World Diary; In Memoriam; Awards and Festivals; and Ivan Butler's thumbnail reviews of the new film books. F. Maurice Speed's *Film Review* is both a tradition and a collector's item.

Jack Oakie's Oakridge by Victoria Horne Oakie. $14.95. (Strawberry Hill Press. Paperback. 126 pages. Illus.)

Victoria Horne Oakie's love for her late husband, comedian Jack Oakie, has become almost obsessive. Some years ago, through Strawberry Hill Press, she published his posthumous autobiography, *Jack Oakie's Double Takes*. It was a charming tribute to a man who was, from my few encounters with him, a very loving human being. Mrs. Oakie's latest tribute is a book devoted to their family home of Oakridge in the Northridge section of the San Fernando Valley (part of the city of Los Angeles). The book includes a considerable number of photographs of the house and its many visitors, all photographed with Oakie. Curiously, many minor or self-proclaimed Hollywood personalities (such as "syndicated columnist" Lee Graham) are scrupulously identified. However, one photograph is captioned only as being of members of the Royal Shakespeare Company. Yet, there in the photograph is a very young Roger Rees, one of the great actors of our time, so memorable in the title role of *Nicholas Nickelby*.

Aside from the photographs, *Jack Oakie's Oakridge* includes brief commentary by Mrs. Oakie, and photographs, magazine articles and other items of Oakie memorabilia. Incidentally, Mrs. Oakie was a talented character actress in her own right, and I wish someday she might consider discussing her own career in print.

The Price Guide to Autographs by George Sanders, Helen Sanders and Ralph Roberts. $21.95. (Wallace-Homestead. Paperback. 608 pages. Index. Illus.)

This is the second edition of an extremely useful guide to values of autographs of everyone from artists and photographers to U. S. Senators and Representatives. The signatures of the last group are probably worth a lot more than they are as a body! The largest number of pages—68—is devoted to entertainers, and the authors provide information as to the value of an individual's signature by itself, on a typed letter or document, on a handwritten letter, and on a photograph. The first part of the book is an introduction to autograph collecting, including essays on the Autopen, the Manuscript Society, and Cartoon Art.

The quoted prices appear to be reasonable, ranging from one dollar for poor Jean Hersholt's signature to $2,200 for that of Greta Garbo. Occasionally, I thought the prices were a little high; for example, surely Lillian Gish has signed and continues to sign so frequently that her signature cannot be worth as much as $25. All in all, this is a welcome and reliable work. Recommended.

Looking into Films: No. 1, RoboCop by H. Arthur Taussig. No Charge. (H. Arthur Taussig. Paperback. 32 pages.)

This is a lengthy analysis of the 1987 feature film, together with a one-page plot summary, a bibliography (unrelated to the film), but no credits. In his introduction, the author notes, "I like science fiction and fantasy films. Psychology and religion are at the center of my analytical interests. I think my essay on *RoboCop* demonstrates both the limitations of my approach and, perhaps, its possibilities." The author also comments, at the end of his essay, "*RoboCop* is not a great movie; perhaps it isn't even a great entertainment," which can only lead the reader to wonder why the author bothered to self-publish this readable yet ultimately unimportant pamphlet.

We Will Always Live in Beverly Hills: Growing Up Crazy in Hollywood by Ned Wynn. $19.95. (William Morrow. Hardcover. 285 pages. Index. Illus.)

Ned Wynn is the son of Keenan Wynn and the grandson of Ed Wynn. His autobiography tells us a great deal about his father and somewhat less about his grandfather, perhaps because he was the more stable member of the family (which is not saying a great deal). *We Will Always Live in Beverly Hills* is a frank admission of the author's growing up into a world of drugs and alcohol, where his friends are the children of the famous or members of pop groups such as the Byrds or the Mamas and the Papas. Much of the book consists, quite obviously, of matters which the author needs to get off his chest, a catharsis of the past.

Openly, Ned Wynn announces that his stepfather, Van Johnson, deserted his mother for the affections of a chorus boy in the London production of *The Music Man.* Obviously he has never forgiven Johnson and his comments on the actor's appearance in later years are somewhat spiteful. In fact, Wynn uses this book to pay off old debts, naming friends, such as Rosalind Russell, who turned against his mother. He paints a vivid and tragic picture of the alcoholics in his life, such as Judy Garland and Keenan Wynn. At the same time, the reader is left with the impression that Ned Wynn has not chosen to reveal everything about his family's past. He was born into a family which consisted of his father, his mother and "Uncle Van," and the true relationship between these three individuals, whatever it may have been, does not emerge from the pages of this book. *We Will Always Live in Beverly Hills* is a volume for those willing to reject the publicized glamor of Hollywood—it is not for the squeamish.

Issues in Feminist Film Criticism edited by Patricia Erens. $45/$17.50. (Indiana University Press. Hardcover/Paperback. 320 pages. Index.)

It is always off-putting to open a book which begins with the dedication, "For Jane, Meryl, Shom, Sondra, and Sue, who taught me what female friendship is all about." Here is perhaps a world in which male readers are not welcome. In 1980, Patricia Erens edited a volume of feminist writings on films titled *Sexual Stratagems: The World of Women and Film.* This current volume continues the cause, collecting together feminist film criticism from the 1980s, with some 29 writers represented. The style and subject matter are varied, ranging from "The Place of Women in the Cinema of Raoul Walsh" to "The Political Aesthetics of the Feminist Documentary Film." A useful bibliography documents feminist film criticism and material on women directors published between 1978 and 1988, and I was pleased that the editor has seen fit to include a glossary for those of us unfamiliar with the cinematic definitions of terms such as "diegesis" or "sadistic scopophilia."

Games of Terror: Halloween, Friday the 13th, and the Films of the Stalker Cycle by Vera Dika. $28.50. (Fairleigh Dickinson University Press. Hardcover. 153 pages. Index.)

Here is the first book to study what the author has designated as "stalker" films, a group of features forming part of the larger "splatter" or "slasher" genres. Concentrating primarily on *Halloween* (1978) and *Friday the 13th* (1979), the author also discusses *Prom Night, Terror Train, Graduation Day, Happy Birthday to Me, Friday the 13th Part 2, Hell Night,* and *The Burning.* Following a brief examination of how each film came into being, the bulk of the author's comments are concerned with the structure of the various features, from a psychoanalytic approach. While some of the analysis is perhaps a little convoluted, I don't believe any reader will have a problem in understanding or enjoying the author's commentary.

While I realize the author is considering the "stalker" films only as a subgenre of the recent "splatter" and "slasher" genres, it might have been interesting to discuss earlier "stalker" films. It is after all a type of feature film which can be traced back to the silent era.

"History of the American Cinema" Series: *Vol. I, The Emergence of Cinema: The American Screen to 1907* by Charles Musser. *Vol. 2, The Transformation of Cinema: 1907-1915* by Eileen Bowser. *Vol. 3, An Evening's Entertainment: The Age of the Silent Feature Picture, 1915-1928* by Richard Koszarski. Each volume $60. (Charles Scribner's Sons. Hardcover. 613 pages/337 pages/395 pages. Index. Illus.)

This project to develop a ten-volume History of the American Cinema has been around for more than a decade. That it has finally, at least in part, come to fruition is for this reviewer nothing short of amazing. That it should have resulted in three such important studies is not only astonishing, but satisfying enough to blunt much of my criticism of the project. Why, therefore, do I want to devote most of this review to a questioning of the series? Simply because I am unhappy with the amount of public money thrown into the project and its elitist, academic slant. The History of the American Cinema series was financed, in substantial part, by the National Endowment for the Humanities. Although there is never any public discussion of how the National Endowment for the Arts or the National Endowment for the Humanities decide on their funding priorities, how their advisors and consultants are appointed, one would like to assume that the National Endowment for the Humanities would have demanded that the series editor have credentials in the field and insisted that the series editorial board consist of prominent writers in the area of American film history. Such is obviously far from the case. The editor whose name appears on the front cover of each book, on the title page, and who has the sole blurb on the back cover of each book is not known in the field. (I have decided he has taken enough publicity for himself on and in the book and I'm not going to name him.) According to the dust jacket blurb, he is the author of *Gradients of Depth in the Cinema Image,* a book with which I am not familiar and which, correct me if I am wrong, is not considered a major work in the history of American film. The editor has also "produced and written an animated segment for *Sesame Street* and a short film on open-heart surgery." Wow! Heavy credentials indeed!

The editor's best credentials, I am reluctant to admit, are that he is an astute businessman, obviously at ease in the highly political world of government funding. I don't know of anyone else who could raise funds to keep a project such as this going for ten years before producing anything. Even the *American Film Institute Catalog* produces volumes faster than this. He is also an editor who obviously does not select his editorial board on the basis of the credentials of the members. On the editorial board, which also receives prominent recognition in each volume, is one Russian (what does he know about American films?), the head of a film archives who has never produced any writing on film history, the coauthor of a book on Japanese cinema, several names of which I have never heard, and only two acknowledged experts in the field, Eileen Bowser and Lewis Jacobs. Kevin Brownlow is not there. William K. Everson is not there. Presumably their academic standards and credentials are considered less than, say, James Leahy, Alan Williams, or John G. Handhardt, members of the editorial board whose names mean nothing in the field of film history.

As for these first three volumes—the remaining seven are announced for publication between fall 1991 and fall 1993—I wonder what is their intended audience. Magnificent as they are, they are too expensive to be textbooks, too expensive for the average film buff, student or scholar. Their uniform, dull grey dust jackets do not attract the casual reader. The third volume with its murky photograph of a film editor hardly evokes "An Evening's Entertainment."

Whoever selected these dust jackets really made a terrible mistake. They cry out "boring." They offer no incentive to open the book. And this is tragic, because each of these three volumes unquestionably offer some of the most important writings to date on the American silent cinema. I don't think one can call them definitive, because I believe the scope of silent film is such that it is impossible to write a definitive volume. Indeed, Richard Koszarski, in his preface, notes how previous writers, including Everson and Brownlow, have studied the period from different viewpoints. (Koszarski considers the filmgoing experience to be of primary importance.)

Koszarski perhaps has the hardest job of all, because he is writing of a period of history already much discussed. Eileen Bowser, writing of the formative years from 1907-1915, takes a revisionist approach, based in part on the findings at an International Federation of Film Archivies (FIAF) conference on early cinema, held in Brighton, England, in 1978. The first volume, which ends in 1907, is by Charles Musser, and it is the most detailed and the longest. Unfortunately, it reads at times rather like a dissertation. In fact, Musser's final paragraph sounds very much like a student's argument for his professor: "I have sought to lay the basis of a new historical perspective for looking at subsequent cinema and even today's moving-image practices in film and video."

Despite the amount of space in the review devoted to criticism of the series, it would be wrong of me to claim anything other than these three volumes are some of the most worthwhile yet written on American silent film. They are compiled by noted authorities on the subject—even if two of the three authors are not considered important enough to serve on the editorial board for this series—and I doubt they will be surpassed, in terms of scholarly standards, by later volumes. As I glance through these books and others on the American silent film I am struck by the thought that no period in American cinema history has been better documented. It seems extraordinary, but it is true. We have the *American Film Institute Catalogs* for the 'teens and the 1920s, we have three volumes from Kevin Brownlow, these three volumes, an ever increasing number of studies of obscure areas, even autobiographies. Has the 1950s cinema received this amount of detailed study and documentation? I don't think so.

Congratulations to Charles Musser, Eileen Bowser, and Richard Koszarski for turning the "History of the American Cinema" series around and proving it to be a worthwhile project. Highly Recommended.

April 1991

Film Study: An Analytical Bibliography by Frank Manchel. 4 volumes. $55 each. (Fairleigh Dickinson University Press. Hardcover. 2,847 pages. Index.)

This is such a monumental work that I have difficulty in describing it. I turn, therefore, to its author, who explains in his Introduction, "This is not a methods book on how to teach about film, though it does offer in each of its seven chapters observations on current practices. Nor is this book an attempt to list definitively the various sources or materials connected with film study, though such a work would be welcome. Rather, it is a survey designed to describe six popular approaches to the study of the cinema, along with a practical analysis of selected books, materials, and information about motion picture rentals."

These areas of study are Film as Film; A Representative Genre of the Film; Stereotyping in Film; A Thematic Approach; Comparative Literature; A Representative Period of American Film (1913-1919); and Approaching the History of Film. Chapters is something of an understatement; for example the last "chapter" makes up an entire volume. Volume 4 of the series is devoted to the index and to a variety of appendices, including selected lists of 16mm, 8mm, and video distributors, the Motion Picture Production Code, a selected list of sources of further film study, the United States economic review of 1984, and the Waldorf Conference statement from 1947.

Each chapter includes a lengthy introduction, detailed reviews of appropriate books (with full bibliographic information), and comments on suggested films with information as to rental sources.

When first published in 1973, *Film Study* was intended to help "lead the way to better and more systematic film programs in colleges and schools." This new edition does far more than that. It provides a unique reference source for documentation on both films and books about them. The book reviews are both informative and entertaining—I don't think I had any disagreement with Manchel's comments, even when he was writing about my own volumes. Film librarians will find these volumes invaluable. Highly Recommended.

Life to Those Shadows by Noël Burch. $45/$19.95. (University of California Press. Hardcover/Paperback. 317 pages. Index. Illus.)

Noël Burch is a writer adored by a small group of academics, but whose name means virtually nothing to the majority of those interested in motion pictures. He is an American who emigrated to France and became a French citizen in 1970. His current book was written in French and subsequently translated into English. It is unfortunate that Burch did not himself write an English-language version, thus obviating the stilted and awkward sentences with which this book is rife.

According to its dust jacket, *Life to Those Shadows* has, as its major thesis, "that film language has a social and economic history, that it evolved in the way it did because of when and where it was constructed—in the capitalist and imperialist west between 1892 and 1929." Oh dear! As readers of my column know by now, any book that throws around words such as capitalist and imperialist as if they have dirty meanings gets short shrift from this reviewer. However, *Life to Those Shadows* is an exception. If you can break through some of the theoretical claptrap, you will find an engrossing and highly personal history of early cinema. British, French, and American films from the preteens are discussed in detail, and the author includes a totally fascinating collection of contemporary quotes on the cinema. I do not pretend to have the intellectual capacity to understand a lot of what Burch writes, but this is a book that I am going to return to again, both for reference and nonassigned reading. Recommended.

Weakness Is a Crime: The Life of Bernarr Macfadden. by Robert Ernst. $34.95/$17.95. (Syracuse University Press. Hardcover/Paperback. 278 pages. Index. Illus.)

Bernarr Macfadden was a health nut in the days before health nuts had any respectability or credibility. He was also publisher of *Photoplay, Movie Weekly, Movie Mirror,* and *Shadowplay.* Unfortunately, the author devotes scant attention—less than a page—to Macfadden's film publications, concentrating on his general readership periodicals, such as *Liberty* and *True Story.* This study of Macfadden's life is rich in facts, but lacklustre in approach. The text is very pedestrian, never really exciting the reader, and, at times, failing to carry a portion of Macfadden's story far enough along. For example, I would have liked to know more about his children, what became of them, are they still alive? Bernarr Macfadden believed in physical fitness, a healthy sex life, appreciation of nudity, and was an avid critic of the American Medical Association. I don't think I can fault him on any of those issues.

Donna Reed: A Bio-Bibliography by Brenda Scott Royce. $35. (Greenwood Press. Hardcover. 143 pages. Index. Illus.)

As might be expected, this volume is similar in format to the other books in the Bio-Bibliography series, containing a brief biographical sketch, a chronology, a filmography, a detailed listing of television and radio appearances, a list of awards and nominations, and an annotated bibliography. There is also a description of the Iowa-based Donna Reed Foundation for the Performing Arts. While, personally, I might ponder the need for such a book on Donna Reed, I was impressed by the amount of information gathered together by the author, particularly for the television and radio listings. The filmography includes full credits for each film, along with a substantial synopsis, quotes from usually favorable reviews and notations as to additional reviews. The style of writing for the 13-page biographical sketch is somewhat terse, but readable, and it is obvious that the author is extremely fond of her subject. Recommended.

And the Winner Is . . . : The History and Politics of the Oscar Awards by Emanuel Levy. $15.95. (A Frederick Ungar Book/Continuum. Paperback. 390 pages. Index. Illus.)

This is the expanded, paperback edition of a book first published in 1987. It is a well-written study of the Oscars, which attempts to find the hidden agendas behind many of the choices. Appendices list the award winners in the six major categories from 1927/28 through 1986, the nationality of the Oscar players (given as a percentage), Best Picture nominees by genre (given as a percentage), Oscar-winning films by number of nominations and commercial appeal, and the most nominated films, 1927-1989.

Frame-Up!: The Untold Story of Roscoe "Fatty" Arbuckle by Andy Edmonds. $19.95. (William Morrow. Hardcover. 335 pages. Index. Illus.)

This new book on Roscoe "Fatty" Arbuckle and his murder trials seems to offer little new to the excellent coverage on the subject provided by David Yallop in his 1976 volume, *The Day the Laughter Stopped.* The author does make the point that Arbuckle was the first person to be the subject of a film industry blacklist. She also offers for the first time what she claims to be "Roscoe Arbuckle's own version of the events of that fateful Labor Day weekend."

As with her previous book on the Thelma Todd murder, *Hot Toddy,* Andy Edmonds has a penchant for quoting conversations to which she was not privy and which were not taped or written down. As the judge at one of Arbuckle's trials might have pointed out, this is nothing more than hearsay. *Frame-Up!* lacks a bibliography, and it also lacks the substance of the David Yallop volume. The author is surprisingly brief in her telling of Arbuckle's last years—perhaps because the comedian's third wife, Addie McPhail, refused to grant her an interview.

Frame-Up! does include a complete and detailed filmography of Arbuckle's work from 1909-1933, compiled by Sam Gill. It is a revised and expanded version of his filmography from the Yallop book.

Produced and Abandoned edited by Michael Sragow. $9.95. (Mercury House. Paperback. 374 pages. Index. Illus.)

Produced and Abandoned is a collection of reviews by members of the National Society of Film Critics of films which failed to receive the attention they deserved on their original release. All of the reviewed films are now available on video (and the video distributors are named). The films are grouped by subject headings devised by the editor. For example, under the heading "Only in America" are such features as *Bound for Glory, Fat City, The Last American Hero,* and *Melvin and Howard. The Right Stuff,* which one might assume belonged in the same category, is reviewed under "Beyond Media Politics," a subject heading which also includes *Salvador* and *Under Fire.* Virtually all of the well-known film critics of the 1990s are represented here, including Pauline Kael, Richard Schickel, Robert Ebert, Judith Crist, and David Ansen. Surprisingly missing is Gene Siskel. This is a book which makes for pleasant, relatively easy reading, but I don't believe the reviews will necessarily encourage anyone to go out and rent the videos.

Hard Core: Power, Pleasure, and the "Frenzy of the Visible" by Linda Williams. $12.95. (University of California Press. Paperback. 342 pages. Index.)

As its title suggests, *Hard Core* is a serious study of film pornography from an intellectual and political viewpoint. The book is at times difficult to read, not because of the language or the subject matter, but because the author demands the total concentration and attention of her readers. She views pornography from a woman's perspective, but does not immediately dismiss it as advocating violence against women, nor does she adopt the slogan, "Pornography is the theory, rape the practice." Curiously, she does not discuss gay pornography as an attraction for women, arguing that she does not wish to address questions raised by a body of films not aimed primarily at her.

Hard Core is the paperback edition of a book first published in 1989.

The Screwball Comedy Films: A History and Filmography, 1934-1942 by Duane Byrge and Robert Milton Miller. $29.95. (McFarland. Hardcover. 155 pages. Index. Illus.)

While there have been a number of books discussing in considerable detail and often in a highly theoretical manner screwball comedies, this is the first volume to provide basic information on the genre. Following a foreword by Arthur Knight (still going and back in the United States after a number of years in Australia), the authors provide brief chapters on major performers, writers and directors of screwball comedies. Then, the authors offer full credits, synopses, and brief critical/historical commentaries on 57 designated screwball comedies from *It Happened One Night* in 1932 through *The Palm Beach Story* in 1942. All the familiar films are here as well as a number of lesser-known titles such as *The Ex-Mrs. Bradford* and *I Met Him in Paris.* The book concludes with a first-rate bibliography and the comments by the authors on the value of the cited works.

This is not, by any means, an analytical study of the genre and the text is somewhat simplistic. However, in this particular case, those are virtues. It is very good to have such a readable, precise survey of one of the more delightful genres in the history of the cinema.

Best American Screenplays 2 edited by Sam Thomas. $35. (Crown. Hardcover. 612 pages.)

Best American Screenplays 2 reprints in their entirety the scripts for *Citizen Kane, Some Like It Hot, The Treasure of Sierra Madre, My Man Godfrey, Judgment at Nuremberg, The Lion in Winter, Terms of Endearment, Julia, Holy Matrimony,* and *The Sting.* Prior to each script, the editor provides full credits and a brief (one or two page) historical note. Useful as is this collection, it would have been helpful had the editor indicated the nature of each script—is it the final shooting script, a cutting continuity or what? However that is a small complaint compared to the value of having such scripts readily available for study and research.

The Bare Facts Video Guide by Craig Hosoda. $9.95. (The Bare Facts. Paperback. 225 pages.)

Previously I reviewed *Movie Buff Checklist,* which documented the nude appearances on screen of various actors. This new volume, *The Bare Facts Video Guide,* takes the process one step further and lists nude appearances by actors and actresses. In fact, the latter get the most attention, and the actors are less covered (no pun intended) than in the earlier work. The author is obviously dedicated to his subject. For each performer he lists major films with dates and characters played. For nude appearances, he indicates the manner of the nudity—frontal, "buns," topless, right breast, left breast, etc.—and provides information as to the number of seconds into the videotape at which the nudity is seen. He even rates the sequences, from "Yawn" (usually too brief or too hard to see) to "Wow!" (don't miss it—the scene usually lasts for a while). There are also references to photographic layouts in periodicals documenting the nude scenes in question. Sorry, there are no illustrations in the book, and because the author is only concerned with films available on videotape he does not necessarily list nude appearances in productions not released on video.

Nudity on screen is obviously an area of study which is growing in appeal. I can anticipate advanced study sessions at some of our more progressive universities. Who knows, in time *The Bare Facts Video Guide* may become a classic text on the subject.

That's Hollywood: A Behind the Scenes Look at 60 of the Greatest Films of All Time by Peter Van Gelder. $22.50. (Harper Collins. Hardcover. 288 pages. Index. Illus.)

That's Hollywood—unrelated to the television series of the same name—takes sixty feature films and provides surprisingly new and original material on each. I was ready to dismiss this as just another general collection of movie information, readily available from a dozen different sources. Such is far from the case.

The author has devised a clever format. Each film is identified by one sentence. For example, *Butch Cassidy and the Sundance Kid* is described as "Two lovable Western outlaws rob one too many trains." the "Talent" (ranging from performers to directors and producers) is similarly described. The text is divided into The Business, Behind the Screen, The Experience, and Can You Spot. . . . The last reveals little known facts or errors in the production. For example, there is one killing every 2.1 minutes in *Rambo: First Blood Part II,* 5 by knifing, 2 by strangulation, 14 by bow and arrow, 15 by firearm, 3 by explosion, and 2 by helicopter assault. Because the book was originally published in England, The Experience often includes comments from British publications. It should also be pointed out that despite the title a number of British feature films (*Brief Encounter, The Lavender Hill Mob,* etc.) are included.

The biggest complaint is that there is no explanation for the choice of films, and it is hard to believe that *Airplane!* or *Carry on Cleo* can be considered among "the greatest films of all time." No silent films are included, and the earliest title is *Frankenstein—Dracula* didn't make it. However, if you want to be entertained, to spend an enjoyable evening reading about a few great and a good many not-so-great movies of the past fifty or so years, this is the book for you. Recommended.

The Hollywood Eye: What Makes Movies Work by Jon Boorstin. $25. (Cornelia & Michael Bessie Books/Harper Collins. Hardcover. 224 pages. Index. Illus.)

Here is a book which is long overdue, a text on motion picture production written by someone in the industry who also understands the aesthetics of filmmaking. The various aspects of production are explained with reference to relevant sequences in popular films from the 1940s through the present. Boorstin points out that we watch films three ways, deriving from them a voyeur's, a vicarious, and a visceral pleasure, and his book examines why every movie is three movies running at once.

The author is easy to understand and his writing style is above reproach. Personal experiences have helped to mold the text, and he writes with both humor and at times pointed accuracy. I wonder how many would have the courage to admit that they do not particularly enjoy watching documentary films but have always wanted to make them. Anyone wanting to understand filmmaking must read this book. It should be required reading matter for every course on film production. Highly Recommended.

May 1991

A Damned Serious Business: My Life in Comedy by Rex Harrison. $21.95. (Bantam Books. Hardcover. 288 pages. Index. Illus.)

Rex Harrison's autobiography is a surprisingly disappointing book. While his comments on some of the people with whom he has worked are interesting and entertaining—I might mention Charles Laughton and Preston Sturges—much of the book appears to have been based on scrapbooks, contemporary reviews, and other materials. Also, Harrison's memory is somewhat selective and flawed. He does not remember how to spell Ronald Colman's name. He makes no reference to any of his films prior to a contract with Alexander Korda. And there is not a hint of his relationship with Carole Landis or her subsequent divorce. The book concludes with an entertaining postscript by Maggie Smith, who does know how to spell Ronald Colman's name, and with a complete listing of Harrison's stage and screen appearances, borrowed from a 1985 biography of the actor by Allen Eyles.

Chains of Gold: Marketing the Ratings and Rating the Markets by Karen S. Buzzard. $29.50. (Scarecrow Press. Hardcover. 226 pages. Index.)

Karen S. Buzzard begins this fascinating, informative, and well-documented history of radio and television ratings with the following delightful comment: "As viewers, we assume that television is about programming, but it is really about audiences—about us. The business of broadcasting is the sale of audiences to advertisers; therefore, as we sit watching the flashing images, all the commercial interests of broadcasting—stations, station representatives, advertisers, agencies, and rating services—sit watching us. It is as though the TV screen were a two-way mirror: we watch from one side; they watch us from the other." Aside from a detailed discussion of the A. C. Nielsen ratings and the other television ratings systems, the author provides the first detailed examination of the pioneering broadcast ratings services, Crossley and Hooper.

The title is taken from a comment by Oren Harris, chairman of the House Committee to Investigate the Ratings: "Broadcasts are bound to ratings with chains of gold which they are reluctant to break."

Bogdanovich's Picture Shows by Thomas J. Harris. $42.50. (Scarecrow Press. Hardcover. 346 pages. Index. Illus.)

Bogdanovich's Picture Shows is a first-rate study of the films of a director who has disappointed more than he has satisfied audiences and critics. The book contains chapters on each of Peter Bogdanovich's films, a chronology, a biographical/career analysis, reviews of books by the director, and a detailed filmography. The analysis of each film is both sensible and readable, and finds negative as well as positive aspects to the director's work. It is easy to dismiss Peter Bogdanovich, but I know from personal experience that he is a fine and decent human being (a man who unlike his Hollywood colleagues does not publicize the goodness of his actions), and based on at least three superb features—*The Last Picture Show, Saint Jack,* and *Mask*—he is a powerful director. Recommended.

The Godfather Companion by Peter Biskind. $8.95. (Harper Collins. Paperback. 193 pages. Illus.)

This book purports to tell "everything you ever wanted to know about all three Godfather movies," and I am sure that it does—if only it was a little easier to use. There is no index, and the "facts" are organized throughout the book under headings such as "Questions: The Book" or "The Numbers" or "How Much Did It Cost." If, like me, you are unfamiliar with the Godfather trilogy, this book is just about useless as a reference source. The most valuable information appears at the back of the volume in the form of complete credits for all three films, vital statistics, ages of the characters, and the Corleone Family Tree.

The Shoot-Em-Ups Ride Again: A Supplement to Shoot-Em-Ups by Buck Rainey. $62.50. (Scarecrow Press. Hardcover. 319 pages. Index. Illus.)

Scarecrow Press seems to have a penchant for publishing books with lousy titles, as evidenced by *The Shoot-Em-Ups Ride Again.* This title is more appropriate for the sequel to a movie about a bunch of drug-crazed hippies. However, despite the modern meaning to shooting up, this is not a volume about heroin addiction but rather a supplement to the author's 1978 work (reprinted by Scarecrow in 1985).

The current volume is divided into five different chapters. Chapter One documents more than 80 films released between 1978 and 1988. Chapter Two covers a little under 500 films released between 1929 and 1978 and not included in the first book, along with 133 European-produced Westerns. Chapter Three contains additional credits for more than 1,500 films from the first volume. Chapter Four documents all Western TV series from 1948-1988. Chapter Five covers Western telefilms.

Each chapter is prefaced by an introductory essay, replete with phrases and expressions which would seem more appropriate to a "B" Western. Credits for the included films are somewhat sparse, limited to production company, running time, players (with no information as to characters played), and in most cases (but not all) director, producer, and screenplay writer(s). Despite some reservations which I have, this is obviously a book with tremendous research value, and one that no Western film buff will want to be without.

Alamo Movies by Frank Thompson. $19.95. (Old Mill Books. Paperback. 128 pages. Index. Illus.)

In his foreword, Fess Parker describes the Alamo as "a treasured American symbol." Author Frank Thompson calls it "a kind of religion." If these are your sentiments, then this is the book for you. *Alamo Movies* is a detailed, primarily factual history of films about the Alamo from *The Immortal Alamo* in 1911 through *Alamo . . . The Price of Freedom* in 1988. Each film warrants a separate chapter, and there are also chapters on the Alamo on Television, Lost Alamo Movies, Alamo Movies That Never Were, and Cameos and Documentaries.

Thompson's facts are considerable and reliable, but his opinions are somewhat questionable. I disagree vehemently with his statement that director Frank Lloyd was "a mediocre talent." The one problem, and a major one, that I had with this book is that I don't have much interest in the Alamo or what went on there, and I cannot get too enthusiastic about a group of films, many of which are undoubtedly pretty awful. I suppose I should get excited reading the "Travis Draws the Line" script extracts accompanying each chapter, but I cannot.

How Green Was My Valley: The Screenplay for the John Ford Directed Film by Philip Dunne. $27.50. (Santa Teresa Press. Hardcover. 105 pages. Illus.)

This book reprints the screenplay for the 1941 20th Century-Fox production of *How Green Was My Valley*, exactly as it was first published by Crown in 1943 in *Twenty Best Film Plays*, edited by John Gassner and Dudley Nichols. Now, as then, there is no indication as to which version this is of the script.

Aside from the reprint of the screenplay, the book includes a 25-page introductory essay by scriptwriter Philip Dunne. It is a detailed account of how the feature came into being, but also contains the usual irritating and often pompous commentary one has come to expect from Dunne. When he begins his essay by attacking the anonymous reviewer in *Time* magazine—could it possibly be James Agree?—who failed to acknowledge Dunne's contribution to the film, one knows that here is a writer with a chip on his shoulder regarding the lack of credit from which screenwriters all too often suffer. I realize that Philip Dunne is a distinguished and much admired writer but he never fails to irritate this reviewer—and for that, perhaps, he deserves praise. (Any book which irritates its reader cannot be totally bad.) Here, he manages to annoy me by criticizing Sara Allgood, who I admire deeply and who, I feel, contributed much to the success of *How Green Was My Valley*. Dunne is dissatisfied with the flat reading of his lines.

Publisher James Pepper has undertaken considerable research to aid Dunne in writing the introduction. He has also put together a beautiful collection of fifteen rare stills, illustrating the text, and produced a book which is a pleasure the handle and peruse. The trade edition is limited to 1,000 copies. I cannot think of any better investment than the limited edition of this work, consisting of 250 copies signed by Dunne and Roddy McDowall, and priced at $35. Despite my personal problems with Philip Dunne's text, I am more than happy to award Highly Recommended status to this volume.

Prime-Time Families: Television Culture in Postwar America by Ella Taylor. $11.95. (University of California Press. Paperback. 208 pages. Index.)

Prime-Time Families examines the changing face of the American family on television from the 1940s through the 1980s, and the influence which the television family has had on the American family. The text is all rather tedious and academic, but presumably the 1989 hardcover edition was popular enough to warrant this present reincarnation in paperback.

The Contemporary Animator by John Halas. $39.95. (Focal Press. Hardcover. 128 pages. Index. Illus.)

John Halas is one of the best-known names in British animation, responsible, with his wife Joy Batchelor, for that country's first animated feature, *Animal Farm*, in 1954. In this profusely illustrated book—more than 70 of the photographs are in color—Halas provides a straightforward, at times almost simplistic, survey of contemporary animation techniques. He discusses computer animation, clay animation, puppet animation, and so forth, mentioning the best-known names in each field, the historical background and the techniques involved. There is a first-rate, nineteen-page glossary of terms, which describes in understandable language everything from A & B roll printing to X-Y-Z axis. A bibliography is also provided. Recommended.

Still in Movement: Shakespeare on Screen by Lorne M. Buchman. $24.95. (Oxford University Press. Hardcover. 179 pages. Index. Illus.)

In her introduction, Lorne M. Buchman writes that the idea of Shakespeare on film conjures the image of Chaplin flustered by the conveyer belt in *Modern Times.* "After four centuries of stage productions of Shakespeare's work, a new machine takes hold of the plays and processes them according to its own particular laws." It's a delightful notion and one that makes a reviewer somewhat more receptive to yet another book on selected film adaptations of Shakespeare's plays. The films which Ms. Buchman has chosen to survey in the relationship between Shakespearean drama and the cinema, representative of the use of cinematic space and time, are all the well-known titles: Laurence Olivier's *Henry V* and *Richard III*, Orson Welles' *Macbeth* and *Othello*, Roman Polanski's *Macbeth*, Grigory Kozintsev's *King Lear*, etc. Apparently there are a growing number of college courses on the subject of Shakespeare on film, and this book will perhaps serve as a textbook for some of them. Unfortunately, it has little interest for most other users.

June 1991

The Soundies Distributing Corporation of America: A History and Filmography by Maurice Terenzio, Scott MacGillivray, and Ted Okuda. $35. (McFarland. Hardcover. 232 pages. Index. Illus.)

Soundies were short, three-minute films of musical numbers or novelty items produced for playing on "Panoram" jukeboxes in the 1940s. They have always held a fascination because of some of the acts which they capture on film, and they are utilized more and more in television compilations because they happen to be in the public domain and are easily accessible. A few years ago, PBS aired a rather poor documentary on the subject, hosted by Cab Calloway.

This new volume is the first (and probably the last) book-length history of Soundies and the company which produced them, and is very welcome for its rich documentation. The authors provide a thorough corporate history of the Soundies Distribution Corporation and list every one of the Soundies, with release dates, technical credits where available, and names of performers. They also list Vis-o-graphs, Featurettes, and Snader Telescriptions, which are often confused with Soundies.

I hate to complain about such a wonderful work of research, but I do have some problems with the book which could have been solved had the authors worked a little harder. (After all, that is what authors are supposed to do according to their readers.) The major source of irritation is the lack of an index to the principal performers in the Soundies. It is impossible without reading through the 151-page filmography to know which Soundies featured, say, Bill Robinson, Lawrence Welk, Smith and Dale, or Eddie Peabody. The index by title is wonderful, but a second index by personality would have been better still. I would have liked to learn something about the actual production of the Soundies, how quickly they were shot, working conditions on the set, etc., and this could perhaps have been provided through quotes from some of the performers in the films. Finally, I would have wished the authors had discussed the racial aspects of Soundies, the fact that Soundies with Black performers were made for showing in Black-only neighborhoods and that the Soundies Distributing Corporation of America even segregated the listings in its catalogs. It's ironic that those Soundies made for Black-only audiences are now probably the most sought-after of all the Soundies, featuring as they do such great stars as Fats Waller, Count Basie, and Duke Ellington.

But please, do not get me wrong. If I was not basically pleased with this book, I would not be devoting so much space to it. I am tremendously impressed by the work of its authors/compilers. Their listings, year by year, are an endless source of information. I did not know that David Raksin (composer of "Laura") had performed in a 1941 Soundie titled "Conga La Maise," or that Alan Ladd is seen in the 1941 Soundie, "I Look at You." Gus Van of Van and Schenck is a very poor performer in Soundies, but I wonder how good were fellow vaudevillians Smith and Dale? I have one of the two Soundies the Duncan Sisters made, but I would love to see the other, "Sweet Onion Time." And I wish somebody would make my day and let me see Mary Brian's Soundie from 1942, "Jealous."

Despite some misgivings, I am happy to rate this book Highly Recommended. I hope its authors will be encouraged to document similar obscure areas of film history.

Hollywood Collectibles by Dian Zillner. $29.95. (Schiffer Publishing. Paperback. 224 pages. Illus.)

Hollywood Collectibles is a delightful book for anyone interested in movie memorabilia. It is divided into sections on Fred Astaire, James Cagney, Charlie Chaplin, Bing Crosby, Dionne Quintuplets, Deanna Durbin, Clark Gable, Judy Garland, Betty Grable, Rita Hayworth, Sonja Henie, Baby Sandy, Al Jolson, Jeanette MacDonald, Charlie McCarthy, Margaret O'Brien, Will Rogers, Shirley Temple, John Wayne, and Jane Withers. Each section includes a lengthy and well-written biography of the performer, with special emphasis given to movie memorabilia generated by his or her screen persona. The illustrations for each section consist of such memorabilia, including books, dolls, lobby cards, sheet music, games, and periodicals. Shirley Temple gets the most coverage—of course her career also extends into politics, and among the memorabilia shown here are items relating to her unsuccessful 1967 campaign for the California 11th district Congressional seat.

The author has also included a list of suggested sources for movie memorabilia, a very useful bibliography, and suggested value ranges for each of the items shown in the book. The publisher has done an excellent job of reproducing the illustrations, although some of the items look a little dog-eared, having been either well used or well loved. Recommended.

Margaret in Hollywood by Darcy O'Brien. $20.95. (William Morrow. Hardcover. 297 pages.)

The author of this novel is the son of George O'Brien and Marguerite Churchill, and a delightful portrait photograph of his mother graces the cover. Despite the title, less than one third of the book is taken up with Margaret's life in Hollywood, with the bulk of the novel devoted to her childhood, the activities of her overambitious stage mother, and the development of the heroine's career as a model and stage actress. The book concludes with Margaret's work in a Western, presumably based on Marguerite Churchill's performance in *The Big Trail.*

This is not a great novel. Darcy O'Brien does not have the distinctive style of, say, Paul Theroux or A. N. Wilson. He writes well and appealingly. Aside from the explicit sex scenes, *Margaret in Hollywood* is in many ways an old-fashioned novel. One enjoys reading the book, but at its close the reader cannot help but wonder what he has gained except for a desire to know how closely this novel is modeled after the career of Marguerite Churchill.

The Federal Communications Commission: A Primer by Robert L. Hilliard. $11.95. (Focal Press. Paperback. 115 pages. Index.)

An independent government agency responsible directly to Congress, the Federal Communications Commission was established primarily to regulate interstate and international communications by radio, television, wire, satellite, and cable. In other words, it is the agency responsible for control of American television. Obviously it is doing a lousy job.

This is a book for anyone (which should mean everybody) interested in the work of the Federal Communications Commission, its background, its politics, and its makeup. In understandable terms, it explains how the Federal Communications Commission functions and provides very detailed background on the agency's history. I found this book extremely helpful and reliable, and the author has done a good job of trying to keep his personal opinion out of the text. Recommended.

The Warrior's Camera: The Cinema of Akira Kurosawa. $39.50/$14.95. (Princeton University Press. Hardcover/Paperback. 366 pages. Index. Illus.)

Akira Kurosawa is the best known of all Japanese filmmakers, with a directorial career extending from 1943 through the present, and, hopefully, long into the future. The current volume examines the development of Kurosawa's style and how it has been used by the director in the making of politically and socially committed films. *The Warrior's Camera* should prove valuable for anyone interested in Japanese filmmaking and is a first-rate textbook for courses devoted to Kurosawa's films.

Monty Pythons Scrapbook/The Life of Brian Script. $13.95. (Heinemann. Paperback. Approx. 128 pages. Illus.)

The Life of Brian, released in 1979, is typical of the Monty Python films in that it is very funny in parts (particularly the Sermon on the Mount sequence), very silly in parts, and often horrendously boring. Unfortunately, the tedious moments are much longer than the naughty bits. The book of the film (I don't really know what to call it because it is a book without an apparent title) includes the complete script, which is as welcome as any film script, and is heavily illustrated.

The reader must turn the volume over and start from the back to read through *Monty Pythons Scrapbook,* which resembles the typical Monty Python production—there are too many boring bits and some of the humor is decidedly childish.

The Life of Brian was the most outrageous of the Monty Python films. It is not a production for the Reverend Donald Wildmon, Senator Jesse Helms, or anyone who is misguided enough to listen to those individuals. Similarly, the book of the film (which was first published in the United Kingdom in 1979 and has just appeared over here) is not for the squeamish. Others can leaf through the pages and stop and enjoy the naughty bits.

Ronald Colman, Gentleman of the Cinema: A Biography and Filmography by R. Dixon Smith. $35. (McFarland. Hardcover. 334 pages. Index. Illus.)

The present volume is an entertaining and informative study of the career of Ronald Colman, combining elements of a biography and a "Films of" compilation. Detailed credits for all of the actor's films are interspersed throughout the text, which is divided into sections: The Image of a Star, Years of Diversification, etc. Each section opens with a general overview of Colman's work during the period, and then the author concentrates on the individual films, offering his own thoughts on the productions, lengthy accounts of the plots, historical background, and occasional quotes from Colman's colleagues. R. Dixon Smith's writing style is similar to Colman's acting—comfortable, reliable, and old-fashioned (and I do not use that term in a derogatory way).

If anything I am surprised that Ronald Colman is popular enough to warrant another book-length study. Of three earlier books, the first was published in 1933 and the most detailed is Lawrence J. Quirk's *The Films of Ronald Colman* (1977). R. Dixon Smith's book certainly supplants all of the previous volumes in terms of depth of study and factual documentation, and as such will be welcomed, but I wonder how many individuals and libraries which have the other volumes will want to spend money for this. Also included is a two-page foreword by Brian Aherne.

Cel Magic: Collecting Animation Art by R. Scott Edwards and Bob Stobener. $19.95. (Laughs Unlimited Incorporated. Paperback. 104 pages. Illus.)

Cel Magic is a long overdue book on what has become an increasingly popular hobby and investment. The work of four animation artists—Bob Clampett, Jay Ward, Walter Lantz, and Shamus Culhane—is discussed, but those chapters are secondary to the coverage on "What Is a Cel?," "How To Buy a Cel," "Trajectory: Market Trends in Animation Art," and "Dealers of Animation." As an added bonus, the authors have included a long and fascinating interview with Dave Smith, founder and curator of the Walt Disney Archives. The book also contains a brief glossary of animation terms and an even briefer bibliography. *Cel Magic* is beautifully produced and well worth owning for the illustrations alone.

I'll Think about That Tomorrow by Evelyn Keyes. $21.95. (Dutton. Hardcover. 340 pages. Illus.)

This is the second autobiographical work from Evelyn Keyes, an actress best known for her roles in *Gone with the Wind, Here Comes Mr. Jordan* and *The Jolson Story.* As might be expected in view of her previous effort, *Scarlett O'Hara's Younger Sister,* Miss Keyes spares neither herself nor those around her in telling of her life after middle age. She writes of her third husband, Artie Shaw, and a relationship which developed from marriage to friendship; of second husband John Huston's last year; of touring with Don Ameche in *No, No Nanette*; of gay young (and not so young) men who make life easier for older actresses; of writing a screenplay based on her first autobiography for Allan Glaser and Tab Hunter, and much, much more. Above all, Miss Keyes writes of her acceptance of advancing years and career changes. She is a woman who does okay for and by herself, and even if you don't care too much about her as an actress, her book is pleasant, relaxed reading.

They Used To Call Me Snow White . . . But I Drifted: Women's Strategic Use of Humor by Regina Barbeca. $18.95. (Viking. Hardcover. 236 pages. Index.)

I like Benny Hill. I published a lengthy interview with him in the ersatz *Playboy* magazine, *Gallery*. I am, therefore, obviously not the right person to be discussing a feminist study of humor in contemporary American culture. Suffice it to say that I did find the book illuminating. Despite being an academic text, it is easy to read, in large part because the author includes sample jokes to emphasize her arguments. As the author intended, I was struck by just how sexist is humor, and how difficult it is for what might be called a "dirty" joke to cross sexual boundaries. The author quotes a very funny joke told against men which involves a certain part of the male anatomy. It's an hilarious story, I laughed aloud at it, but there is no way that I can ever repeat it and nor I doubt can any other man.

International Film Prizes: An Encyclopedia by Tad Bentley Hammer. $95. (Garland. Hardcover. 922 pages. Illus.)

International Film Prizes is one of the most useful reference books to come along in quite some time. As its title suggests, this vast work provides listings of film-related awards in 42 countries, plus the European Film Awards, from Argentina through Yugoslavia. Not only are complete listings given, but also the author provides a brief history of each award. I know from my own research just how difficult it is to find this type of information, and I have nothing but admiration for the author's skill in ferreting out factual data on, say, the Crystal Star (from France) or the Hong Kong Film Awards.

The most sought-after information will probably relate to film awards from the United Kingdom and the United States. For the former, the Stella/Bafta Awards are given, as well as the British Film Critics' Circle Awards, the Evening Standard British Film Awards, and the Sutherland Trophy/British Film Institute Award. The Academy Awards lead off the American section, and the author has included not only all winners in all categories, but also nominees for the top seven Oscars. Also provided are the National Board of Review winners (now known as the D. W. Griffith Awards), the New York Film Critics Circle Awards, the National Society of Film Critics' Awards, the Los Angeles Film Critics' Awards, the Golden Globes, and the Independent Spirit Awards.

As added bonuses, the book includes title and personality indexes, thus making it possible to determine how many awards a film or an individual have received worldwide. This is a superb achievement in compilation. Perhaps it is a little expensive for the individual purchaser, but no library can, or should be without it. Highly Recommended.

Before Caligari: German Cinema 1895-1920 edited by Paolo Cherchi Usai and Lorenzo Codelli. $75. (Edizioni Biblioteca dell'Immagine, distributed in the United States by the University of Wisconsin Press. Hardcover. 533 pages. Index. Illus.)

Before Caligari is an important collection of essays on German cinema prior to 1920. Included are chapters on the arrival of the Cinématographe Lumière in Germany in 1896, the origins of official film censorship, film and German colonial propaganda for the Black African territories, music in German cinema prior to 1918, emergence of the feature film between 1910 and 1911, Max Reinhardt and film 1910-1914, Asta Nielsen's early films, the Skladanowsky brothers, and much, much more. There is also a complete listing of German fictional films from 1895-1920 surviving in archives belonging to the International Federation of Film Archives, but, of course, the researcher is up against the secretive attitude of that organization and there is no indication as to which archives has which film.

Before Caligari is published in both Italian and English, with the latter language occupying the right-hand pages. This is another volume made possible thanks to the efforts of Italy's Pordenone Silent Film Festival and is, of course, recommended.

The Films of Freddie Francis by Wheeler Winston Dixon. $37.50. (Scarecrow Press. Hardcover. 315 pages. Index. Illus.)

The Films of Freddie Francis, No. 24 in the Filmmakers Series, concentrates on the subject's work as a director, but also includes much information on his film career as a cinematographer. The book contains two interviews with Francis, an interview with Peter Cushing, a detailed filmography with critical commentary on each of Francis's films as a director, a comprehensive bibliography, and a listing of reviews of directed films.

Music of the Shadows: The Use of Musical Accompaniment with Silent Films, 1896-1936 by David Robinson. (Le Giornate del Cinema Muto. Paperback. 112 pages. Illus.)

Music of the Shadows was published as a catalog to an exhibit at the 1990 Pordenone Silent Film Festival. It is beautifully illustrated and contains some useful information, but should not be considered a definitive reference work on the subject (which is in need of a book-length study). I spotted at least one outrageous error and that is the author's claim that Joseph Carl Breil's score was used to accompany *The Birth of a Nation* at Clune's Auditorium in Los Angeles. It was, of course, Carli Elinor who both composed the score and conducted the orchestra for the original Los Angeles performances of *The Birth.* Breil's score was used at the Riverside, California, previews and the New York opening. The text of *Music of the Shadows* is in both Italian and English.

Naming Names by Victor S. Navasky. $9.95. (Penguin Books. Paperback. 515 pages. Index.)

Naming Names is a major study of the activities of the House Committee on Un-American Activities and its investigation into Communism in the Hollywood film industry in the late 1940s and 1950s, first published in 1980 and winner of the 1981 American Book Award. It is published here in paperback for the first time. The author concentrates on the Committee's use of informers who named names and in so doing blackened not only their own names but also the American political system. *Naming Names* is fascinating, readable, and frightening.

Famous Movie Detectives II by Michael R. Pitts. $47.50. (Scarecrow Press. Hardcover. 357 pages. Index. Illus.)

This is a supplemental volume to Michael Pitts' first book on the subject, published in 1979. It contains chapters on Arsene Lupin, Hercule Poirot, Inspector Clouseau, Inspector Maigret, Mike Hammer, Miss Jane Marple, Nurse Sarah Keate, Perry Mason, Philip Marlowe, Raffles, Sexton Blake, The Shadow, Sherlock Holmes, and

The Whistler, with each chapter concluding with a filmography. Also included are brief essays and filmographies on "Other Movie Detectives": Bill and Sally Reardon, C. Auguste Dupin, Duncan MacLain, Father Brown, Flash Casey, Frank Cannon, Hank Hyer, Inspector Hornleigh, Jack Packard and Doc Long, Joe Dancer, Joel and Garda Shane, Kitty O'Day, Kojak, Lew Archer, Lord Peter Wimsey, Philip Trent, The Roving Reporters, Russ Ashton, Shaft, Thatcher Colt, Tony Rome, Travis McGee, and Wally Benton. There is also a listing of novels featuring the detectives discussed in the book and additions and corrections to the base volume. A useful work from the always reliable Michael Pitts.

July 1991

Science Fiction Stars and Horror Heroes: Interviews with Actors, Directors, Producers and Writers of the 1940s through 1960s by Tom Weaver. $35. (McFarland. Hardcover. 461 pages. Index. Illus.)

Tom Weaver continues the work he began with *Interviews with B Science Fiction and Horror Movie Makers* (1988) and interviews a second batch of 28 actors, writers, directors, and producers associated with the science fiction and horror genres: Acquanetta, Phyllis Coates, Hazel Court, Richard Devon, Gene Fowler, Jr., Arthur Gardner, Arnold Laven, Paul Landres, Pat Fielder, Albert Glasser, Bernard Glasser, Herk Harvey, Gordon Hessler, Louis M. Heyward, John Howard, Kim Hunter, Robert Hutton, Nancy Kovack, Anna Lee, Janet Leigh, Richard Matheson, Gregg Palmer, Mala Powers, Robert Shayne, Yvette Vickers, Katherine Victor, Virgil W. Vogel, and Robb White.

One does have the feeling that Weaver is scraping the bottom of the barrel with this book, but the scrapings are not without interest. It is particularly good to read interviews with those not primarily associated with horror or science fiction, such as Anna Lee, who has long needed a detailed survey of her career. (Incidentally, I doubt she was ever known as "Queen of the Quota Quickies," and quite a few of her British films do not fall into that category.) Each of the interviews is wrapped around with a short introductory comment and a basic filmography with film title, producer/distributor, and year of release. Many of the interviews have previously been published, in abridged form, in buff publications, such as *Fangoria* and *Starlog*.

Movie History: A Survey by Douglas Gomery. $33.75. (Wadsworth. Paperback. 526 pages. Index. Illus.)

Obviously intended primarily as a text book, and a first-rate one at that, *Movie History: A Survey* provides a concise overview of film history from 1895 to the present. The book is divided into three sections: The Silent Cinema: 1895-1927, The Hollywood Studio Era: 1928-1950 and The Television Era: 1951-1975. While the American film industry is the author's main concern, filmmaking in Europe, the Soviet Union, and Japan is not ignored. The book's value is enhanced by a helpful guide to further reading and filmographies on the major figures discussed in the text.

Douglas Gomery is one of the best academics in the film field, and he also reviews all the new film books in *Communication Booknotes*. This is the perfect volume for anyone needing access to a compact, reliable, and readable history of the cinema, and is recommended.

Variety, 1990: The Year in Review edited by Marilyn J. Matelski. $21.95. (Focal Press. Paperback. 182 pages. Index.)

This is the second year that prominent articles and statistical information have been reprinted in book form from the trade publication, *Variety*. Items are grouped in the following categories: Film, Homevideo, Music-Records, and Theatre. The book also includes a brief history of *Variety* (exactly as it appeared the previous year), an excruciatingly boring piece on Research Methodologies in Mass Communication, an index to 1990 *Variety* film reviews, and a necrology.

In that the book is concerned primarily with documentation, the reprinted reviews tend to be heavy on facts, but rather dull reading matter. Unfortunately none of the entertaining pieces from *Variety* are included. Nevertheless this is an important volume, well worth owning as a continuing source of information. Recommended.

Boris Karloff: A Critical Account of His Screen, Stage, Radio, Television, and Recording Work by Scott Allen Nollen. $39.95. (McFarland. Hardcover. 486 pages. Index. Illus.)

This is at least the seventh book on Boris Karloff, which must make the actor one of the most written about in film history. It is a little difficult to justify yet another book on Karloff, but in the author's defense it must be noted that he has produced a detailed critical biography, which examines each of Karloff's starring vehicles. According to the publicity release from the publisher, author Nollen spent almost fifteen years in researching the volume—presumably not full-time—and, from a personal viewpoint, it is obvious that a great deal of love and devotion has gone into the project. There are comments from Karloff's colleagues, including Henry Brandon, Vincent Price, Zita Johann, and David Manners, as well as quotes from contemporary sources and from the dialogue of Karloff's films, all of which enliven a text which is readable and authoritative.

The area which is the least detailed, and which is always glossed over in the Karloff books, is the early period in which he toured with various stock companies and had "bit" parts in a number of films. We always read that Karloff made his screen debut in *The Dumb Girl of Portici,* but has anyone yet examined this extant film and tried to identify Karloff in the cast? I notice that before coming to California, Karloff was in Chicago, and it makes me wonder could he possibly have been in the portion of the film shot there rather than at Universal? These are questions which surely a Karloff biographer could answer.

In addition to a first-rate text, the author includes a detailed filmography, a detailed list of television and radio performances, a detailed listing of Broadway appearances and a not-so-detailed list of stock company activities, a list of recordings, and information on proposed film roles. Recommended.

The Complete James Bond Movie Encyclopedia by Steven Jay Rubin. $27.95. (Contemporary Books. Hardcover. 475 pages. Illus.)

Steven Jay Rubin is the author of *The James Bond Films,* and obviously knows everything there is to know about the subject. To prove it, he has compiled *The Complete James Bond Movie Encyclopedia,* which contains entries on just every Bondian subject one can imagine, and quite a few one cannot. There are entries on the obvious, including George Baker and Roger Moore, and the obscure, such as Blayden Safe House and Hyper-Intensified Magnetic Field. Entries on the various Bond productions include complete credits, and I doubt there is a Bond actor or character which has been ignored. There appears to be at least one error: if the author was English he would know that Kent is not a British city, but an English county.

This is the reference book which every librarian who has ever had to answer a question on the James Bond films will welcome with open arms. It is also the perfect gift for any Bond enthusiast. I only hope there will be more James Bond features so that Steven Jay Rubin, who seems to be a one-subject author, can continue with his obsession. Recommended.

The Official Abbott & Costello Scrapbook by Stephen Cox and John Lofflin. $13.95. (Contemporary Books. Paperback. 224 pages. Illus.)

While intended primarily for Abbott and Costello fans, this book does include a considerable amount of useful documentation, including credits, synopses, and production sidelights on all of the comedic pair's feature films. There is also an intelligent and well-considered introductory essay on the art of Abbott and Costello.

The Official Abbott & Costello Scrapbook includes a foreword by Buddy Hackett (who played Lou Costello in a TV movie). The book is printed on what appears to be the cheapest possible paper—I don't know if this is supposed to tie in with the somewhat cheap qualities of Abbott and Costello's work. Despite its title, there is no indication anywhere in the book as to why this if the *Official* scrapbook.

Hollywood Be Thy Name: Random Recollections of a Movie Veteran from Silents to Talkies to TV by William Bakewell. $35. (Scarecrow Press. Hardcover. 264 pages. Index. Illus.)

Hollywood Be Thy Name, No. 25 in the Filmmakers series, is the autobiography of veteran Hollywood actor William Bakewell, whose film career began in 1925 with *A Regular Fellow.* Bakewell discusses his career, the colleagues with whom he worked (notably Lew Ayres and Clark Gable), and the Motion Picture and Television Fund, of whose board he is a member. *Hollywood Be Thy Name* also includes a filmography prepared by Michael Hawks.

The Missing Reel: The Untold Story of the Lost Inventor of Moving Pictures by Christopher Rawlence. $19.95. (Atheneum. Hardcover. 306 pages. Index. Illus.)

Augustin Le Prince was one of the pioneers in the development of the motion picture. He vanished after boarding a train from Dijon to Paris on September 16, 1890. In 1976, Christopher Rawlence purchased Le Prince's former family home in the Northern England city of Leeds, totally unaware of the work of the inventor. A few years later, finding his home in danger of demolition by the city council, Rawlence began an effort to save it and have it declared an historic monument by researching Le Prince's life and work.

The end result of Rawlence's considerable efforts is a documentary (which I have not seen) and this book. I don't know and Rawlence does not reveal if the house was saved from the wrecking ball. *The Missing Reel* moves backwards and forwards in time from the present to the last century. Rawlence discusses his contemporary research activities as he examines the work of Le Prince, Thomas Edison, and Eadweard Muybridge, along with what happened to Le Prince's wife and family after the inventor's disappearance. The result is a fascinating book, which not only illuminates film history but also reveals the work and travel necessary to document a period which is inadequately covered in contemporary film texts.

Unfortunately, Rawlence does not uncover what happened to Le Prince on that fateful day in September 1890, and that is a basic problem with *The Missing Reel.* The reader feels unsatisfied. What happened to Le Prince? Was he murdered by a rival inventor? Unlikely. Did he have a secret liaison (a gay lover perhaps which is suggested and rejected by Rawlence) which necessitated his leaving his wife and family? Equally unlikely. Did he choose to commit suicide? Perhaps the most plausible solution. But if so, why when he was so close to perfecting his camera/projector? And why did he leave no note, and why did he choose such a curious manner in which to disappear? Obviously no one will ever know what happened to Augustin Le Prince.

Rawlence's book provides us at least with a detailed history of Le Prince's life up to September 1890, and fully documents his prominent place in the invention of the motion picture. Recommended.

To Wit: Skin and Bones of Comedy by Penelope Gilliatt. $24.95. (Charles Scribner's Sons. Hardcover. 327 pages. Index. Illus.)

Penelope Gilliatt presents us with an entertaining and intelligent overview of comedy in its many shades and hues. The former film critic of *The New Yorker* might seem an unlikely choice to author such a book, but her photograph on the back cover reveals a woman with a puckish face who might well have a view on life similar to Whoopi Goldberg (who is one of the performers profiled in the book). After quoting Goethe, "to explain is to destroy" (a one-liner which most film academics would do well to memorize), Gilliatt is careful not to harm her subject by overexamination. Her thoughts on individual comics and films range from one paragraph to several pages. Her skill is in linkage—she discusses Buster Keaton and Samuel Beckett together, Charlie Chaplin and James Thurber, Spike Milligan and Robin Williams, Elaine May in *Enter Laughing* and Beatrice Lillie in *Exit Smiling.*

Some of the subjects may not be familiar to American readers—Charlie Drake, Morecambe and Wise, the brilliant Mrs. Shufflewick—but certainly they should be. Some of the comedy may be illusive—how many would expect to come across an examination of *The Sound of Music* ("the most awful musical we have seen since the war") in a book on the skin and bones of comedy. Some of the included comedians may have been overdiscussed—but who has written more intelligently on John Cleese than Penelope Gilliatt? Recommended.

Resisting Images: Essays on Cinema and History edited by Robert Sklar and Charles Musser. $44.95. (Temple University Press. Hardcover. 343 pages. Illus.)

This collection of essays is one in the series of Radical History Review Books, and it is apparently the radical nature of the subjects covered which links the various essays. That, plus, the editors point out, the questions about film history within the context of recent debates over historiography and cultural theory which the articles address. Such statements might be enough to put off most readers, but, happily, a few of the included essays are of some interest, although the radical aspects of their subjects are somewhat questionable. Coeditor Charles Musser writes well on "Work, Ideology, and Chaplin's Tramp." But multimillionaire, womanizing Charlie Chaplin can hardly be considered a radical figure in film history, J. Edgar Hoover's views notwithstanding.

Stephen J. Ross's "Cinema and Class Conflict: Labor, Capital, the State, and American Silent Film" is an important study, but it was published only last year in another book, and it is doubtful there is a need for another reprint so soon. Coeditor Robert Sklar's essay, "Oh! Allthusser!: Historiography and the Rise of Cinema Studies" is pompously academic, self-serving, and virtually unreadable. Other essays cover such topics as Cinema and Everyday Life: Soviet Worker Clubs in the 1920s; Ideology Elsewhere: Contradictory Models of Italian Fascist Cinema; The Female Spy: Gender, Nationality, and War in *I See a Dark Stranger;* and An "Other" History: The New Latin American Cinema. Most are readable and authoritative.

As should be apparent, there is little linking these essays, and it might have been better had the editors decided to concentrate on a specific area of film history rather than offer such a vague perspective on world cinema. The failure of editors and publisher to include an index is indefensible in a supposed academic text.

Hollywood Songsters: A Biographical Dictionary by James Robert Parish and Michael R. Pitts. $75. (Garland. Hardcover. 842 pages. Index.)

Parish and Pitts have selected 104 film performers who are also singers and for each provide an enthusiastic biographical essay, a title filmography, listings of radio, television and stage appearances, and an album discography. The selected entertainers cover the entire history of the sound film from the late 1920s (Al Jolson and Sophie Tucker) to the present (Madonna and Bette Midler). All types of singers are also included from the operatic (Lily Pons) to legends of rock 'n' roll (Elvis Presley). If the entries have faults, it is that they are all overlaudatory. Ann Sothern has an "abundance of talent, versatility, and beauty." Mickey Rooney is "irrepressible." Barbra Streisand is "an enduring superstar." This is an expensive but worthwhile reference work. And for those who complain that most academic publishers, such as Garland, Scarecrow, and McFarland, do not include dust jackets with their books, Garland has taken an obvious, and welcome, step in printing a photograph (of Ethel Merman) directly on to the cover of *Hollywood Songsters*. It's a neat way to make an academic book more appealing, and other publishers should take note.

Hollywood Beauty: Linda Darnell and the American Dream by Ronald L. Davis. $19.95. (University of Oklahoma Press. Hardcover. 231 pages. Index. Illus.)

Hollywood Beauty is a straightforward, somewhat dull text on the life and career of Linda Darnell, written by an academic (a professor of history at Southern Methodist University), who is obviously more than a little besotted with his subject. His admiration for Darnell has led him to overstate her importance—he calls her life an American tragedy, "like Judy Garland's and Marilyn Monroe's and James Dean's." And, of course, Linda Darnell cannot possibly be considered in the same category of stardom or acting talent.

This is a rather remarkable book in that it is published by a University Press, which would not, one assumes, normally consider works of this nature, and it is presented as an academic text (albeit an understandable one) with detailed notes as to sources. There does not seem to have been a man or woman who knew or worked with Darnell whom the author has not interviewed, although the text does not indicate that such interviews were of much importance in the final analysis.

If you like Linda Darnell, you will like *Hollywood Beauty*. If you have no strong feelings about the actress, you will quickly become tired of reading about her greatness and the tragedies of her life, such as being forced to move from Bel Air to Sherman Oaks. *Hollywood Beauty* also includes a detailed filmography and listings of television and stage appearances.

August 1991

The summer is always the slow season for new books. This month's column is, therefore, devoted to reviews of film books from 1989 which went unnoticed at the time.

The Hidden Cinema: British Film Censorship in Action, 1913-1972 by James C. Robertson. $58. (Routledge, Chapman & Hall. Hardcover. 190 pages. Index. Illus.)

Through the first major examination of the records of the British Board of Film Censors, the author explains how film censorship works in the United Kingdom. He deals with films on a case-by-case basis, in chronological order, discussing their historical context and the reaction of the British Board of Film Censors to them. It is fascinating to discover that Theda Bara's first film, *A Fool There Was,* was the only one starring the actress to be rejected outright, and it has, apparently, never been publicly shown in the United Kingdom. Or to learn that because of a ban on the materialization of Christ on the screen, distributors of *King of Kings* did not even bother to present the film for approval to the British Board of Film Censors but went to individual local authorities for permission to screen.

Schickel on Film: Encounters—Critical and Personal—with Movie Immortals by Richard Schickel. $20.95. (William Morrow. Hardcover. 312 pages. Index.)

This book collects Schickel's previously published essays on Charles Chaplin, John Ford, Preston Sturges, Alfred Hitchcock, Stanley Kubrick, The Hollywood Ten, Ronald Reagan, Douglas Fairbanks, Sr., Harold Lloyd, Gary Cooper, James Cagney, Humphrey Bogart, Marlon Brando, and Woody Allen. Schickel is surprisingly outspoken in his comments on The Hollywood Ten and in his objection to the Hollywood Left's entitlement "to command the historical record unchallenged." In fact, while he is far from my favorite critic, one must admire Schickel for writing what he thinks without regard to popular opinion.

The Illustrated History of the Soviet Cinema by Neya Zorkaya. $25. (Hippocrene Books. Hardcover. 320 pages. Illus.)

Despite its title, this is not a collection of photographs but rather a general historical survey of Russian filmmaking from the 1890s through the late 1980s. Published by arrangement with the Novosti Press Publishing House in Moscow, the book might be read as an official Soviet view of filmmaking, but it appeared after Glasnost and is, therefore, surprisingly frank and honest. The author, who is a leading Soviet film critic, even points out that the official Soviet filmmaking body, Goskino, implies "obstacles, barriers and pressure."

Moving Pictures by Anne Hollander. $29.95. (Alfred A. Knopf. Hardcover. 512 pages. Index. Illus.)

In her introduction, Anne Hollander notes that "Movies, like paintings, are intended for all time," unlike theatre which is ephemeral. She links the tradition of European painting to modern cinema, suggesting how the two are interrelated. For example, Hollander compares Dreyer's *Day of Wrath* and Milestone's *The Front Page* with Rembrandt's "Syndics of the Draper's Guild." However, the majority of the text and the illustrations relate not to the art of the motion picture but to the art of painting.

Retakes: Behind the Scenes of 500 Classic Movies by John Eastman. $4.95. (Ballantine Books. Paperback. 388 pages.)

Self-described as "a delightful pop history of the movies," *Retakes* begins with *A Nous la Liberté* (1931) and ends with *Zulu* (1964), proving that the word "classic" is a highly subjective adjective. For each title, the author includes dates and place where filmed, leading players and director, and brief production information. The last is the type of material usually to be found in press releases, although in his introduction the author maintains he has avoided such items as "notoriously inaccurate and unreliable." The choice of films is based on video availability, those rated best by critical, popular, and industry standards (I didn't know the film industry had any standards), and those films on which behind-the-scenes information was available. The book makes easy if not inspired reading.

Landscaping the American Dream: The Gardens and Film Sets of Florence Yoch, 1890-1972 by James J. Yoch. $45. (Harry N. Abrams. Hardcover. 243 pages. Index. Illus.)

Landscaping the American Dream is one of the most interesting and unusual film books of recent years, although it is probably of more appeal to gardening enthusiasts than film buffs. With 200 illustrations, it documents the career of landscape architect Florence Yoch, including the gardens she designed for Jack Warner, Dorothy Arzner, and David O. Selznick, and the movie sets which she landscaped, beginning in 1935 with *The Garden of Allah*, and including *Romeo and Juliet, The Good Earth, Gone with the Wind*, and *How Green Was My Valley*.

On the Other Hand: A Life Story by Fay Wray. $16.95. (St. Martin's Press. Hardcover. 270 pages. Index. Illus.)

Anyone who has the honor to know Fay Wray would expect her autobiography to be an intelligent and sensible appraisal of her life and of those around her. She does not disappoint. She writes with frankness and honesty of her films and of her relationships with four writers, two of whom she married (John Monk Saunders and Robert Riskin) and two she did not (Clifford Odets and Sinclair Lewis). The filmography which accompanies the text begins with the 1925 feature, *The Coast Patrol*, produced by Bud Barsky (who wanted to change her name to June Darling). After that it could only get better.

The Wizard of Oz: The Official 50th Anniversary Pictorial History by John Fricke, Jay Scarfone, and William Stillman. $29.95. (Warner Books. Hardcover. 246 pages. Illus.)

Here is a beautifully produced, profusely illustrated history of the popular M-G-M musical, with chapters on the film's creation, promotion, reception, merchandizing, television airings, etc., ending with its recognition as an icon. The book also includes a short bibliography and an introduction by Jack Haley, Jr. What is particularly pleasing about this work is that despite the considerable amount of material already published on *The Wizard of Oz,* the authors have managed to come up with a surprisingly large number of new illustrations.

The West That Never Was by Tony Thomas. $14.95. (Citadel Press. Paperback. 225 pages. Illus.)

Based on the PBS series of the same name, the title is perhaps a little misleading. This is not a comparative study of the Hollywood Western against the real West, but a series of informative essays on sixteen "B" Westerns and 24 "A" Westerns, from *Rider of Death Valley* in 1932 to *The Shootist* in 1976. Brief credits for each film are also included, and the foreword features a short interview with Tim McCoy, whom Thomas describes as "a far more erudite and well-spoken man than his image as a movie cowboy would have led most people to believe."

Louise Brooks by Barry Paris. $24.95. (Alfred A. Knopf. Hardcover. 609 pages. Index. Illus.)

Thanks to a handful of devoted and influential fans, Louise Brooks became a legendary silent star. In reality, she was never more than an adequate actress who never achieved stardom and who, in later years, carefully nurtured fables about her career and the careers of those around her, helping to give the impression that her films were far more important than they really were. Neither the life nor the career warrant such a scholarly, detailed, and meticulous biography as this. It is unfortunate that one so minor should be the subject of a text so major.

When in Disgrace by Budd Boetticher. $35. (Neville Publishers, Inc. Hardcover. 408 pages. Illus.)

This fascinating and readable autobiography does not tell too much about Boetticher's screen career, but does provide vivid portraits of his personal life, bullfighting, and years in Mexico (including a lengthy section on life in a Mexican jail). Many of the film-related items in *When in Disgrace* are short, to-the-point and always new and revealing, such as the half-page word portrait of Laird Cregar announcing "I am . . . queer" to the cast of *Blood and Sand*, or Boetticher's comments on Gilbert Roland's "Being a major pain in everyone's ass" during the filming of *Bullfighter and the Lady*.

When in Disgrace also includes a brief preface by Robert Stack, a filmography by Chris Wicking and an afterword by William K. Everson. It does not include a much-needed index. The first edition is limited to 1,000 copies.

September 1991

Felix: The Twisted Tale of the World's Most Famous Cat by John Canemaker. $30. (Pantheon Books. Hardcover. 188 pages. Index. Illus.)

Animation artist and historian John Canemaker has compiled a magnificent study of the creation and continued marketing of Felix the Cat, one of the best-known figures in the history of American animation. Profusely illustrated with black-and-white and color photographs of scenes, toys, newspaper cartoons, and documents, this book is everything a film book should be (and all too seldom is). It is a credit to both its author and, for the quality of the production, its publisher. It includes a selected filmography beginning with *Feline Follies* in 1919 and concluding with *Felix the Cat: The Movie* in 1989. It also contains a bibliography and, joy of joys, a flickbook cartoon of Felix the Cat, which readers can enjoy by flipping through the right hand pages of the book.

Felix is a continuing source of wonderment and delight. On one page is Ann Pennington teaching the Black Bottom to Felix. On another, we are reminded that in 1928 NBC tested the first RCA television camera using a rotating Felix the Cat doll. Long may Felix keep on walking and John Canemaker continue documenting animation history. Highly Recommended.

From Agent to Actor: An Unsentimental Education or What the Other Half Knows by Edgar Small. $14.95. (Samuel French. Paperback. 245 pages.)

Edgar Small began his career, as an actor, in the 1940s, became a producer in the early 1950s, and took up employment as an agent at the end of the decade. His book is not really an autobiography, but more a first-hand account of the relationship between actors and agents. It will obviously be of tremendous interest to anyone starting out in the acting profession and wanting to understand more about what an agent can do for them and what an agent, in turn, will require of them. (Some sample Screen Actors Guild contracts are included in an appendix.)

From Agent to Actor is also a book with more than its share of interesting anecdotes. For example, Yul Brynner was the first actor (in 1956) to demand that the morals clause be dropped (by Paramount) from his contract. Three actors—Walter Hampden, Alan Napier, and Cedric Hardwicke—all played the same role in the 1956 Lana Turner vehicle, *Diane,* all are in the film, all are a different height, and no one in the filmgoing audience seemed to care. As Small points out, it proves that no actor is irreplaceable.

Abbott and Costello in Hollywood by Bob Furmanek and Ron Palumbo. $16.95. (A Perigree Book. Paperback. 271 pages. Index. Illus.)

Previously I reviewed *The Official Abbott & Costello Scrapbook.* This new volume is infinitely superior, and with its introduction by the children of Abbott and Costello is obviously far more deserving of the title "official."

In its opening chapters, authors Furmanek and Palumbo discuss the work of the comedians prior to Hollywood and provide a general commentary on the making of the movies. What follows is about as good a production survey as it is possible to compile. For each of Abbott and Costello's films, the authors provide complete credits (and I do mean complete), information on production start and finish, copyright date, release date, date of the first script draft. Next, the authors offer a synopsis, background information, production notes, discussions of the cast, of Abbott and Costello's scenes, of the critical reaction to the film, and, finally, a postscript. The book concludes with a brief commentary on Abbott and Costello after they split up and a reprint of "Who's on First" as heard in *The Gay Nineties.*

I could have done without the advertisement for MCA/Universal's release of the comedians' films. Advertisements do not belong in books. Except for the aforementioned item, this is about as good a "Films of" book as it is possible to compile. It could not be better. It is so good that I can even tolerate the introduction by Jerry Lewis. Highly Recommended.

The Third World in Film and Video, 1984-1990 by Helen W. Cyr. $29.50. (Scarecrow Press. Hardcover, 256 pages. Index.)

The third in a series of volumes, which began in 1976, *The Third World in Film and Video* provides brief entries on more than 1,100 films available in 16mm and video. For each film, the author lists producer, distributor, running time, format, and a brief (sometimes only one line) summary. The author is extremely generous in her use of the term "Third World," allowing it to include such obviously pro-Western nations as Japan, Hong Kong, and Singapore. While useful to educators, books such as *The Third World in Film and Video* will soon become obsolete with the demise of nontheatrical film and increasingly detailed listings of video releases now available.

D. W. Griffith and the Origins of American Narrative Film by Tom Gunning. $42.50. (University of Illinois Press. Hardcover. 326 pages. Index. Illus.)

In large part because almost all the films survive and there is considerable contemporary documentation, the short subjects directed by D. W. Griffith at American Biograph are probably the most documented and discussed in all of film history. Each new book provides further insights and thoughts on the subject. The latest, Tom Gunning's *D. W. Griffith and the Origins of American Narrative Film,* is no exception. Although scholarly in approach, happily it is both readable and understandable, providing the user with an intelligent overview of the change in film style and content from 1900 through 1915, with the author concentrating on the years 1908 and 1909 (which Gunning considers crucial in Griffith's and the cinema's development).

The book includes a solid bibliography and an impressive list of source materials. It is illustrated with frame enlargements from the films discussed. The book's jacket is graced by a photograph which I have never before seen of Griffith directing Blanche Sweet in a scene from *Judith of Bethulia.* Recommended.

No Turn Unstoned: The Worst Ever Theatrical Reviews by Diana Rigg. $13.95. (Silman-James Press/Samuel French. Paperback. 192 pages. Index.)

Of Diana Rigg's nude scene in the 1970 New York production of *Abelard and Heloise,* critic John Simon wrote, "Diana Rigg is built like a brick mausoleum with insufficient flying buttresses." That is a sample of the many hundreds of negative comments gathered together by Miss Rigg, best known these days as the host of the PBS series, *Mystery!* This is a delightful book for browsing although, surprisingly, quite a few of the reviews are rather tedious and, I am sure, far more dull than the productions about which their authors are writing.

The Best of Universal by Tony Thomas. $9.95. (Vestal Press. 102 pages. Index. Illus.)

This is a brief, popular history of the Universal, with chapters devoted to specific genres: classic, Western, musical, romance, horror, and comedy. As is always true with Tony Thomas, the text is easy to read, but there is probably little new here for film buffs. Also, because the book is sanctioned by MCA/Universal, do not expect any negative criticism of the studio's product.

Unspeakable Images: Ethnicity and the American Cinema edited by Lester D. Friedman. $45/$18.95. (University of Illinois Press. Hardcover/Paperback. 455 pages. Illus.)

Unspeakable Images is a collection of essays on ethnicity in American films, discussing such issues as Eddie Cantor and *Whoopee,*

Catholicism in the films of John Ford, Cecil B. DeMille and *The Cheat,* and Hollywood's image of Latin America. Depending on one's viewpoint, the essays are either tedious or sophisticated. What always surprises me with regard to academic books such as this is the amount of time and assistance the editor seems to need to produce the book (sabbatical leave and eight individuals helping with everything from proofreading to "technical and emotional assistance"), and the failure of the editor and publisher to realize the need for an index.

The Moguls: Hollywood's Merchants of Myth by Norman Zierold. $14.95. (Silman-James Press/Samuel French. Paperback. 354 pages. Index. Illus.)

This is a welcome reprint of Norman Zierold's 1969 book on the major Hollywood moguls: David O. Selznick, Carl Laemmle, Samuel Goldwyn, B. P. Schulberg, Cecil B. DeMille, Adolph Zukor, Harry Cohn, the Warner Bros., and Darryl F. Zanuck, In the intervening years since the first publication of this book, biographies of a number of the subjects have appeared (notably Goldwyn), but *The Moguls* remains a useful overview of the men who created the Hollywood studio system.

The Sir Harry Lauder Discography by Darrell Baker and Larry F. Kiner. $42.50. (Scarecrow Press. Hardcover. 222 pages. Index. Illus.)

As its title implies, *The Sir Harry Lauder Discography* provides a complete and detailed listing of every recording by the Scottish-born music hall and vaudeville entertainer. The earliest is "To Jericho," recorded in February 1902, and the last is not a record at all but a BBC radio broadcast on December 25, 1942, of Lauder's best-known song, "I Love a Lassie." The inclusion of the latter is indicative of the content of this book, which is far more than a mere discography. It contains also a biographical sketch, a composite compilation of interviews in which Lauder discusses the writing of his songs, a complete listing of such songs, a discussion of Lauder's films with a filmography, listings of cylinder recordings, 78 rpm recordings, cassette tapes, and long-playing records, and a bibliography.

There are at least two errors—the suggestion that two of Lauder's films were shot in 16mm—but otherwise this appears to be a definitive volume on Lauder and one that anyone interested in his career will wish to own. Unfortunately, I suspect that such are few in number and that is regrettable in view of the performer's once proud status as a folk hero of two continents. Recommended.

The Columbia Checklist: The Feature Films, Serials, Cartoons and Short Subjects of Columbia Pictures Corporation, 1922-1988 by Len D. Martin. $65. (McFarland. Hardcover. 646 pages. Index.)

A long-needed reference work, *The Columbia Checklist* purports to list all the films produced and/or distributed by the Hollywood company which grew from poverty row to major studio status. For each feature film, the author provides release date, running time, genre classification, director, producer, script credits, and leading players, together with a short plot summary. Much the same information is provided for the serials, while lesser credits and no plot summaries are offered for the cartoons (1928-1965), Miscellaneous Short Subjects (1922-1965), and Independent Short Subjects (1959-1971). Appendices document the Columbia Movie Series, the Columbia Westerns and Stars, the Columbia Comedy Shorts and Stars, and Columbia Academy Award winners and nominees. A name index is particularly valuable.

The only problem, and it is a minor one, is that the compiler has chosen to include non-Columbia, foreign-produced films which the studio released in the United States. As a result, it is a little puzzling to come across, for example, British International Pictures' *Abdul the Damned*, listed as a Columbia feature. A small complaint for a Recommended book.

The Films of the Eighties: A Complete, Qualitative Filmography to Over 3400 Feature-Length English Language Films, Theatrical and Video-Only, Released Between January 1, 1980 and December 31, 1989 by Robert A. Nowlan and Gwendolyn Wright Nowlan. $68.50. (McFarland. Hardcover. 867 pages. Index.)

The Films of the Eighties provides entries on English language films from the United States, the United Kingdom, Australia, Canada, and New Zealand, released in the 1980s. Each entry includes information on the production or releasing company, producer, director, screenwriter, composer, cinematographer, editor, and occasional additional credits, together with running time, year of release, and leading players with characters portrayed.

In that it is unlikely that the American Film Institute Catalog will ever get around to documenting the 1980s, the present volume might seem worthwhile. Certainly it is a useful, quick reference guide for basic documentation. But unfortunately it has many negative characteristics, notably the insistence of the authors in including not simply a plot summary but a critical commentary in which is included the basic plot. Such commentaries are far from adequate, and the opinions of the authors and their friends are far removed from serious criticism. There are occasions when the commentaries suggest the authors have not seen the films about which they are anxious to offer opinions. Plot summaries included in these commentaries are sometimes inaccurate.

In that most films of the 1980s and 1990s have multiple production companies, it is hard to accept the simplicity of the credits given here in that category—and is it the production company or the distributor listed? *A Bigger Splash* from 1974 is included because the authors claim it was put on the shelf until 1984—yet I saw it at a commercial movie house in 1974. Using reverse logic, the authors do include *Brenda Starr,* made in 1989, but unreleased as of 1991!

It is typical of the inaccuracy and valueless commentary in place of plot summary that the authors devote seven lines in the entry on *The Whales of August* asking if it is not about time the Academy of Motion Picture Arts and Sciences gave Lillian Gish an Honorary Oscar. Sorry to tell you this, authors, but the Academy did—twenty years ago. Yes, a tremendous amount of work must have gone into compiling this volume. Yes, it has some value. But it has too many flaws to be acceptable as a reference work. A far better purchase would be a complete set of *Monthly Film Bulletin* from the 1980s.

A Heart at Fire's Center: The Life and Music of Bernard Herrmann by Steven C. Smith. $29.95. (University of California Press. Hardcover. 425 pages. Index. Illus.)

Bernard Herrmann began his career as a film composer with *Citizen Kane* in 1941—"it's been a downhill run ever since"—and for the next thirty years contributed scores to such films as *The Magnificent Ambersons, The Day the Earth Stood Still, The Man in the Gray Flannel Suit, Vertigo, Psycho,* and *Taxi Driver.* Herrmann will always be closely linked to the careers of Orson Welles (with whom he first worked in radio in 1938) and Alfred Hitchcock. He was uncompromising and difficult in both his professional and his personal life. As author Steven C. Smith points out, "Herrmann committed himself to a creed of personal integrity at the price of unpopularity."

A Heart at Fire's Center is equally uncompromising in its portrait of one of the few legends of the film music world. With passion, vigor, and an unswerving concern for accuracy, Steven C. Smith discusses both Herrmann's life and his work, which, of course, encompassed far more than the composition of film scores. A detailed appendix lists the composer's work in film, the concert hall, radio, and the recording studio. Steven C. Smith is a music critic who understands music. Equally important, he understands how to entertain and hold the reader. Recommended.

The Nostalgia Entertainment Sourcebook compiled by Randy Skretvedt and Jordan R. Young. $19.95/$9.95. (Moonstone Press. Hardcover/Paperback. 158 pages. Index.)

The Nostalgia Entertainment Sourcebook is an extremely useful compilation of addresses in the categories of Music, Movies, Radio, and Theatre. Included are film and video dealers, fan clubs, periodicals, record companies, sheet music sources, research centers, and much, much more. The information is very basic, but hard to locate in one sourcebook. No library or film buff should be without it. Recommended.

Incidentally, Skretvedt and Young are coeditors of the quarterly *Past Times: The Nostalgia Entertainment Newsletter,* also published by the Moonstone Press, and which I am happy also to recommend.

Five for Five: The Films of Spike Lee by Terry McMillan, Toni Cade Bambara, Nelson George, Charles Johnson, Henry Louis Gates, Jr.,

Melvin Van Peebles, with photographs by David Lee. $24.95. (Stewart, Tabori & Chang. Paperback. 215 pages. Illus.)

Five for Five is a beautifully produced volume of photographs by David Lee of scenes and production shots from the five feature films of his much-vaunted brother (in both the literal and spiritual sense) Spike Lee: *She's Gotta Have It, School Daze, Do the Right Things, Mo' Better Blues,* and *Jungle Fever: From Harlem to Bensonhurst and Back.* Each of the films is lauded in an essay by a different writer, and some of the text is perhaps a little too enthusiastic. (*Mo' Better Blues,* for example, is compared, favorably, to *Citizen Kane.*) And Melvin Van Peebles contributes (in the words of the publisher) "a joyful foreword." Interestingly, it is not Spike Lee who is the star of this book, but David Lee's photographs, which would stand out in their own right regardless of the films which they are promoting.

Backstory 2: Interviews with Screenwriters of the 1940s and 1950s edited by Pat McGilligan. $29.95. (University of California Press. Hardcover. 425 pages. Index. Illus.)

Backstory 2 continues the series of interviews which began with the 1986 publication of *Backstory: Interviews with Screenwriters of Hollywood's Golden Age.* Included in the present volume are Leigh Brackett, Richard Brooks, Betty Comden and Adolph Green, Garson Kanin, Dorothy Kingsley, Arthur Laurents, Ben Maddow, Daniel Mainwaring, Walter Reisch, Curt Siodmak, Stewart Stern, Daniel Taradash, and Philip Yordan. Aside from the fascinating interviews, the book also includes brief introductory essays on each individual and basic filmographies. Even if you are not particularly interested in the career of the individuals concerned, the interviews still make intriguing reading. My favorites were Dorothy Kingsley and Philip Yordan. Comden and Green tried too hard to be entertaining, and Garson Kanin was, well, Garson Kanin. Recommended.

Hollywood in Friuli: Sul Set di "Addio alle Armi" by Carlo Gaberscek and Livio Jacob. (La Cinteca del Friuli. Hardcover. 263 pages. Illus.)

I have noticed an interesting phenomenon with regard to European publication of film-related books. The British produce a great many volumes, the majority of which are superficial, lacking in solid research and documentation, and generally available in the United States only as "remainders." The Italians publish fewer books, but they are models of research and scholarship, beautifully produced, and generally on neglected topics.

What prompts my comment is the publication of *Hollywood in Friuli,* a detailed account of the filming of *A Farewell to Arms* in the medieval town of Venzone and the Friuli district of Italy. The authors have gathered together just about every possible material on the filming of this epic, which was the last of its type to use Italian locations and the last production of David O. Selznick. The volume is lavishly produced and includes a bibliography which must surely list just about every newspaper and periodical reference to the production. Recommended.

Jean Renoir: A Life in Pictures by Célia Bertin. $29.95. (The Johns Hopkins University Press. Hardcover. 403 pages. Index. Illus.)

Despite its title, this is not a picture book on the great French and international film director, Jean Renoir, but rather a detailed and readable account of his life and work, translated from the French by Mireille Muellner and Leonard Muellner. Sanctioned by Renoir's widow and son, Célia Bertin's book is a charming remembrance of a charming yet determined director who changed the face of world cinema. It also contains a detailed filmography with plot summaries for each of Renoir's films.

Before the Nickelodeon: Edwin S. Porter and the Edison Manufacturing Company by Charles Musser. $60/$29.95. (University of California Press. Hardcover/Paperback. 603 pages. Index. Illus.)

This is a successor to Charles Musser's documentary film of the same name on the same subject. In fact, there is an unintentionally amusing photograph on the book's dust jacket purportedly showing Musser directing his narrator (unidentified as such) Blanche Sweet. The direction appears to consist of Musser's Scotch-taping a page of the script. I don't think Blanche was that taken with the film, as she told me she could neither remember Edwin S. Porter's name nor the correct title of his 1903 film, *The Great Train Robbery.*

New York University professor Robert Sklar, who is not identified as coeditor on a recent film book with Musser, is quoted on the dust jacket that this is "the most important book on American early cinema yet to appear." In that *Before the Nickelodeon* deals only with a specific area of film history I would not agree with Sklar, but there is no question that this is an important text and, undoubtedly, the definitive volume on the subject of Edwin S. Porter's work with the Edison Manufacturing Company. Unfortunately, the text is ponderous, and will serve more for reference purposes than leisurely reading. It is also surprising that Musser virtually ignores Porter's career after Edison, devoting only a few pages to it, and, apparently, not even bothering to view the extant films from that later period.

Before the Nickelodeon is the first volume in the UCLA Film and Television Archive Studies in History, Criticism, and Theory. It is a curious choice in that Porter and his films are closely associated with the Museum of Modern Art, and I do not believe any of his films are housed at UCLA. The Archive connection also results in an introduction by that organization's director which I cannot resist from quoting:

"Musser rejects timeworn and reductive explanations of historical causality such as auterism and technological determinism. His dialectical approach to historic narrative freely crosses disciplinary boundaries to convey the complex interplay of aesthetics with a multifaceted social and industrial context."

So there!

You'll Never Eat Lunch in This Town Again by Julia Phillips. $22. (Random House. Hardcover. 591 pages. Index.)

Julia Phillips details the problems in her life and career as the first female producer to win an Oscar for Best Picture. The problems range from cocaine addiction to concern that her office must have an adjoining bathroom (to serve as a retreat when snorting cocaine), a combination of which there is apparently a paucity at Ray Stark's company. *You'll Never Eat Lunch in This Town Again* is not a nice book. Written in a chatty style it is an exposé of Hollywood in the 1980s, a confirmation that the American film industry really is as sleazy as most people believe it to be. If nothing else, the book is a refreshing breath of reality compared to the artificial view of contemporary Hollywood presented by such apologists as *Entertainment Tonight.* I'll take Julia Phillips over Mary Hart any day.

Film and Propaganda in America: A Documentary History, Volume IV, 1945 and After edited by Lawrence H. Suid. $75. (Greenwood Press. Hardcover. 536 pages. Index.)

Reprinted here (in their original format) are government-related documents on the *Why We Fight* series for Cold War audiences; American film policy in Occupied Germany; the Atomic/Soviet threat in 1950; Cold War recruiting films; military support for Hollywood feature films, film censorship, and film's impact in the 1950s; John Kennedy, *PT-109,* and the USIA; and Hollywood, the Pentagon, and Vietnam.

Of course, quite a few of the items make dull reading, but on the whole this is a fascinating collection of materials. It's great fun to read John Wayne's letter to the president, announcing his intention to film *The Green Berets,* and Jack Valenti's memorandum to the president that "Wayne's politics are wrong." Who would have thought that the U. S. Marine Corps could expend so much time and effort in worrying about the length of Michael Conrad's hair in *Baby Blue Marine.* Or that the White House would concern itself with the possibility that Cliff Robertson might portray John F. Kennedy smoking a cigarette in a long white cigarette holder.

There is quite obviously a lot more to government than one might imagine from reading the newspapers. As with the other volumes in this series, here is an important primary documentation source for historians, scholars, and researchers.

October 1991

In and Out of Synch: The Awakening of a Cine-Dreamer by Noël Burch. $54.95. (Gower Publishing Company. Hardcover. 254 pages. Index. Illus.)

Gathered together here are a collection of essays by the well-known theorist and critic Noël Burch, written between 1970 and 1985. They cover such diverse topics as Fritz Lang, early American cinema and the avant garde, Marcel L'Herbier, Sergei M. Eisenstein, Carl Theodor Dreyer, and Edwin S. Porter. As with all of Burch's writings, these reprinted pieces are both irritating and thought-provoking. The former emotion derives from an early realization that Burch is often ignorant of historical facts familiar to most film buffs. The enjoyment and stimulation in the essays is from Burch's original dissection and discussion of his subjects. Whatever one's opinion of Burch—and no one obviously has an opinion of Burch higher than Burch himself—he is certainly never boring.

The Elvis Files: Was His Death Faked? by Gail Brewer-Giorgio. $16.95. (Shapolsky Publishers. Hardcover. 275 pages. Illus.)

Author Gail Brewer-Giorgio has spent ten years investigating if Elvis Presley is still alive. Her confirmation that his death was a hoax is based on such revelatory evidence as no one has claimed his multimillion dollar life insurance policy, pictures of his mother have been mysteriously removed from Graceland, and (truly mind-boggling) Elvis' name is misspelled on his tombstone. If you want to be well and truly dead make certain the mortuary gets your name right!

Perhaps the most stunning piece of evidence and the one which this whole book into perspective is a photograph on the dust jacket showing Elvis watching mourners from the Graceland poolhouse. I have looked and looked and looked at this photograph—it is of a door and no one is there.

Osgood and Anthony Perkins by Laura Kay Palmer. $45. (McFarland. Hardcover. 423 pages. Index. Illus.)

I cannot honestly confess that I have felt there is much of a need for a book on either Osgood Perkins or his son Anthony, but now that Laura Kay Palmer has compiled such a work I can only congratulate her for her dedication and attention to detail. In strict chronological order, Ms. Palmer provides detailed listings for each actor's work on stage, screen, radio, and television. There are full credits and commentary for each production in whatever medium. The amount of information which Ms. Palmer has been able to gather on Osgood Perkins' work in the early Film Guild and *Chronicles of America* films is evidence of the amount of research she has undertaken. (Incidentally, recently I saw *Second Fiddle,* produced by the Film Guild in 1923, and I did not recognize Osgood Perkins.)

The author is quite obviously devoted to her subjects and permits personal biases to intrude at times, but there is nothing objectionable in this. The book is, incidentally, an unauthorized study—Anthony Perkins was unwilling to cooperate on the project. Rather than deny a book such as this, he should be pleased that anyone would consider his career worthy of attention. Recommended.

Scenes Unseen: Unreleased and Uncompleted Films from the World's Master Filmmakers, 1912-1990 by Harry Waldman. $35. (McFarland. Hardcover. 285 pages. Index. Illus.)

Harry Waldman has come up with a highly original idea, documenting film projects which might have been from directors and actors. He divides his book into sections titled after the reasons why the films were not made, such as "The Producer's Veto," "On the Wrong Side of the War," or "Too Hot to Handle." Within each section are chapters devoted to individuals and the films they were prevented from completing. For example, under Josef von Sternberg we find *I Claudius* and a produced but lost film, *A Woman of the Sea.* Under Errol Flynn is listed *William Tell,* under George Cukor *Something's Got to Give,* and under Henry Hathaway *I Loved a Soldier.* Despite my examples of American filmmakers, the bulk of the book is devoted to European cinema and directors such as Abel Gance, Paul Fejos, Robert Bresson, Thorold Dickinson, and Mauritz Stiller. The author provides whatever credits are known for the unrealized projects and offers a factual commentary on the history of each film and its undoing. Recommended.

More of Hollywood's Unsolved Mysteries by John Austin. $18.95. (Shapolsky Publishers. Hardcover. 268 pages. Illus.)

More of Hollywood's Unsolved Mysteries offers gossipy examinations of the deaths of Elvis Presley (or nondeath), Thomas H. Ince, Thelma Todd, Carol Wayne, Carole Landis, Marilyn Monroe, Paul Bern, Lupe Velez, and Nick Adams. The entire book is rather "tacky" but great fun to read, full of hyperbole such as "Why Did Lupe Refuse To Have an Abortion" or "What Happened to Nick's Tape Recorder." The author concludes by asking, "Just Who Was Johnny Roselli," and then links the gangster to the deaths of Todd, Bern, Jean Harlow, Vicki Morgan, and Marilyn Monroe. If nothing else, Austin deserves praise for exposing the hypocrisy of the Hollywood press which seeks to suppress rather than publish the truth about the film industry.

Repeat Performances: A Guide to Hollywood Movie Remakes by Doris Milberg. $12.95. (Broadway Press. Paperback. 230 pages. Index. Illus.)

Repeat Performances does not provide much more information than is not already known by most film buffs. The compiler lists a film, with date of release, production company, director, and two or three leading players, followed by its remake, with the same credit information. Television productions are included as well as theatrical features. A couple of lines of basic information follows each entry together with a one-line plot summary. Because the entries are limited to only one original and one remake, additional remakes or earlier, less popular versions are mentioned either in the notes or totally ignored (this is particularly true of silent versions).

Films are not listed in straight alphabetical order, but are grouped under categories such as mysteries, musicals, or classics (although the compiler uses more "cutesy" headings). To find a specific film, it is necessary first to turn to the index which lists both original and remake titles.

Collector's Guide to Treasures from the Silver Screen by John Hegenberger. $16.95. (Wallace-Homestead. Paperback. 232 pages. Illus.)

Lots of illustrations but a slight text are features of this latest book on collecting movie memorabilia. The author provides some useful names and addresses but far better listings are available in *The Nostal-*

gia Entertainment Sourcebook (reviewed last previously) and my own *Collector's Guide to Movie Memorabilia* (now out of print) provides a factual and accurate overview of the subject.

Fly Away Home: Memoirs of a Hollywood Stunt Man by John Weld. $19.95. (Mission Publishing. Hardcover. 249 pages. Illus.)

Despite its title, John Weld's autobiography does not have a great deal to do with Hollywood, and will perhaps be of more interest to those familiar with his work as a novelist. (I regret that I am not among that number.) Weld writes of working as a stuntman in the 1920s (doubling for Earle Foxe, ZaSu Pitts, and others), an early friendship with Clark Gable while the latter was still an extra, writing the screenplay for *The Invisible Man* and failing to receive credit, a love affair with Walter Huston's wife Nan Sutherland, handling publicity at Warner Bros. in the mid 1930s, and of a roommate named Steve Runyon who had an affair with Jean Harlow as part of his job as a studio publicist and whose disappearance led to Howard Strickling's meeting his wife-to-be. The Runyon story is interesting. *The Invisible Man* saga left the reader wanting to know more, particularly how a novelist as respected as R. C. Sheriff would take credit for another man's work.

It is all moderately informative and entertaining, as is the telling of the rest of Weld's life. The book is written by a man of obvious intelligence and capability as a writer—he worked with legendary editor Maxwell Perkins—and yet it failed to evince much enthusiasm from this reader. Somehow one has the sense of a lack of substance.

The Fleischer Story by Leslie Cabarga. $16.95. (Da Capo. Paperback. 216 pages. Illus.)

Leslie Cabarga has compiled a first-rate history of the Fleischer Cartoon Studio from 1920-1942, documenting its production of the animated shorts of Betty Boop, Popeye, Koko the Clown, and Bimbo, and giving what its author considers to be long overdue credit to Max Fleischer's brother Dave. (In the introduction, Cabarga writes of the reaction of Max Fleischer's family to his telling them that he felt Dave had not received the credit he deserved. It's a fascinating part of his story, and I would have welcomed more on this matter.) *The Fleischer Story* is well written and researched and profusely illustrated. Cabarga designed and laid out the book himself, and so here is one volume for which an author can legitimately take full credit. A title filmography, listing all productions from 1916-1944 broken down by series (Screen Songs, Cartoons, Superman, etc.) is also included. On the negative side, there is no index, no bibliography and the lack of footnotes documenting sources is a disappointment. However, the paperback edition of *The Fleischer Story* is more than welcome and Recommended.

The BFI Companion to the Western edited by Edward Buscombe. $26.95. (Da Capo. Paperback. 432 pages. Illus.)

An institutional publication is always suspect, and I greeted the paperback edition of *The BFI Companion to the Western*—the hardcover edition was published in 1988—with suspicion. My suspicions were further aroused by the opening words in the foreword by rent-a-critic Richard Schickel, "An Encyclopedia of Western Movies! I never thought I'd live to see the day." Why, Richard, why?

Happily I had the courage to progress further and discovered that this is a superb compilation on the Western genre with just about every entry one can imagine. I would have preferred a straight alphabetical order, but can certainly live with the entries grouped alphabetically in four major subgroups: A Cultural and Historical Dictionary, A Select Guide to the Western Film, Dictionary of Western Film-Makers, and Television Westerns. It is good to find entries on everything and everybody from Posses and the Reno Gang to *Ramrod* and Clayton Moore. Of course, there are errors and omissions. No book of this magnitude could be with or without them. For example, the Television Westerns should have included the networks on which the series first aired. The book also includes a short (40-page) history of the Western; a chart documenting the years in which the various Western heroes, from G. M. Anderson to Clint Eastwood, worked; a table of Westerns year by year, compared by percentage to other releases; and an annotated bibliography which is remarkably fair and honest in its appraisals. Highly Recommended.

The Disney Touch: How a Daring Management Team Revived an Entertainment Empire by Ron Grover. $22.95. (Business One Irwin. Hardcover. 334 pages. Index. Illus.)

Some of the best—perhaps the best—articles on the Hollywood film industry as an industry rather than an entertainment conglomerate appear in the pages of *Business Week*. And it is that magazine's Los Angeles Bureau Chief who is responsible for this solidly researched and readable history of the Walt Disney organization, beginning in 1984 when Michael Eisner and Frank Wells were hired as its corporate leaders. "Once a lacklustre theme park company, the Walt Disney Company of the 1990s is an entertainment giant," writes Grover, noting the organization's expansion into the production of what is best described perhaps as semiadult entertainment with Touchstone and Hollywood Pictures, its acquisition of Jack Wrather's Disneyland Hotel and the *Spruce Goose* and *Queen Mary* attractions in Long Beach, and the plans to build a theme park in France. If this was about any American business, it would be fascinating reading; in that the book is about the Disney empire of the 1980s and 1990s, it is doubly enjoyable. Recommended.

The Gorgon's Gaze: German Cinema, Expressionism, and the Image of Horror by Paul Coates. $49.50. (Cambridge University Press. Hardcover. 301 pages. Index.)

Intended obviously as a textbook, *The Gorgon's Gaze* is a study of recurrent themes in German cinema from the 1920s onwards, offering no new film discoveries and concentrating on the same "old" titles from the same respectable directors. Highly academic, I would dismiss *The Gorgon's Gaze* as totally unreadable, but, intentionally or not, its author does tend to draw some amusing conclusions from his analyses. My favorite comment relates to Alfred Hitchcock's use in his films of knives, which the author considers to be weapons of castration. On page 82, he makes this extraordinary statement regarding Hitchcock, "known to legend as 'Hitch,' the fat castrato whose partial loss of name corresponds to loss of the part of his anatomy designated by it."

Other than for outrageous comments such as that, *The Gorgon's Gaze* is not worthy of attention. As with so many of these academic film texts, it drops names like "Metz" and "Lacan," which mean nothing to more than 99.9% of the American public, and yet are thrown around as household words which do not even require use of their first names.

Hitchcock's Rereleased Films: From Rope to Vertigo edited by Walter Raubicheck and Walter Srebnick. $39.95/$17.95. (Wayne State University Press. Hardcover/Paperback. 302 pages. Index. Illus.)

Could it possibly be that the more prestigious the university the more esoteric and unreadable the books published by it press? I might believe so if it were not for the entertainment-oriented volumes from the Oxford University Press, which are neither esoteric nor unreadable, simply not very good. From Cambridge University, I turn to Wayne State University Press (and what a nice refreshing change of pace), which has published a group of seven titles in its "Contemporary Film and Television" series. The most recent considers the five Alfred Hitchcock features reissued in 1983-1984: *Rope, Rear Window, The Trouble with Harry, The Man Who Knew Too Much* (1956), and *Vertigo.*

The editors of *Hitchcock's Rereleased Films* have gathered together essays by fifteen "academics," including such relatively well-known names as John Belton and Robin Wood, together with a "talk" by *Vertigo*'s screenwriter Samuel Taylor. I am happy to report that virtually all of the pieces are readable and interesting, and while not all are totally original in thought or concept, many offer new insights into the director's work.

The Films of Carlos Saura: The Practice of Seeing. by Marvin D'Lugo. $45/$14.95. (Princeton University Press. Hardcover/Paperback. 263 pages. Index. Illus.)

Carlos Saura is the best known of contemporary Spanish filmmakers and one whom I have admired since viewing *Cria Cuervos* in 1976, a film which stars Saura's wife and Chaplin's daughter Geraldine. (I should qualify my opening statement and write that I admire everything in the film except for one embarrassing frontal nude shot of Ms. Chaplin.) The current offering is welcome as the first book-length, English language study of Saura's films from *Los Golfos* in 1959 through *The Dark Night* in 1989. Each of the films is discussed in considerable details, with the author providing some historical background but generally concentrating on an auterist analysis of the production. I would have welcomed a filmography.

Funny Woman: The Life and Times of Fanny Brice by Barbara W. Grossman. $35. (Indiana University Press. Hardcover. 301 pages. Index. Illus.)

For those who were not around to see her on stage or hear her as "Baby Snooks" on radio, it is perhaps a little difficult to appreciate the importance of Fanny Brice in the history of 20th century American entertainment. In agreement with her current biographer, I don't feel that Barbra Streisand did her justice; Streisand has neither the warmth nor the comedic genius of Brice.

Surprisingly, apart from a 1953 biography published shortly after her death, there has been no book-length study of Fanny Brice until the present. That omission is well filled by Barbara W. Grossman's superbly researched and eminently readable book, based primarily on contemporary sources. Brice's family did little to cooperate—although the author was able to talk on the telephone to the entertainer's son and daughter—and so Ms. Grossman has relied on the written word, undertaking, for example, the mammoth task of reading *Variety*, cover to cover, from 1905 until Brice's obituary was published there 46 years later. Grossman's book makes fascinating reading, answering for once and for all the riddle of how Miss Brice wished her first name to be written (she used Fanny rather than Fannie), comparing discrepancies in Brice's reminiscences with the reality of her early years, and reporting the extraordinary story that son-in-law Ray Stark was so upset by the Norman Katkov biography that he paid $50,000 to have the printing plates destroyed.

In addition to providing an excellent bibliography, Ms. Grossman has documented all U. S. sources of information on Brice. Perhaps a filmography and a listing of stage appearances would have been nice, but that is a minor quibble regarding a book which is Highly Recommended.

November 1991

Gordon MacRae: A Bio-Bibliography by Bruce R. Leiby. $42.95. (Greenwood Press. Hardcover. 242 pages. Index. Illus.)

While I realize that MacRae's fans will disagree with me, I find it difficult to understand the need for a reference source for the singer/actor such as this. I am sorry but I can think of dozens and dozens of performers far more deserving of this type of documentation. Having made that comment, I must applaud Bruce R. Leiby for his work. Following a succinct if somewhat terse biographical essay (which makes no attempt to hide MacRae's alcoholism), Leiby provides detailed listings of the performer's work on stage, radio, screen, television, and in night clubs and on the concert stage. He also includes a discography, a listing of sheet music associated with MacRae and an annotated bibliography. Certainly MacRae's fans and his family should be grateful to Bruce R. Leiby. I doubt there is an aspect of MacRae's public life which is not fully documented.

Prison Pictures from Hollywood: Plots, Critiques, Casts and Credits for 293 Theatrical and Made-for-Television Releases by James Robert Parish. $49.95. (McFarland. Hardcover. 543 pages. Index. Illus.)

The blurb accompanying this book indicates that James Robert Parish is the author of over 75 books on the entertainment industry. I am sure that must be an understatement as I seem to have received at least that number of books for review in the past three years. Certainly it seems that way!

The content of Parish's present offering is described in the book's subtitle. It is the usual Parish tried-and-true formula, with no surprises and little if anything to criticize. Credits for each production are detailed, and Parish provides short essays which offer the reader a plot summary, an occasional quote from the dialogue, and a sampling of contemporary reviews. Parish is generally reliable in that he has obviously viewed most of the films on videotape. Where he does make errors is in his commentary on some of the films which probably he has not seen; for example, his summary of *The Godless Girl* is not completely correct and he does not indicate that Fritz Feld, and not Cecil B. DeMille, directed the dialogue sequence.

Films are arranged in alphabetical order, from *The Accusing Finger* (1936) to *You Only Live Once* (1937). A one-page listing of Prison series on television, compiled by Vincent Terrace, is included, and a complete index to all personalities mentioned in the text and essays is particularly valuable.

Life Is Too Short by Mickey Rooney. $22.50. (Villard Books. Hardcover. 384 pages. Index. Illus.)

Mickey Rooney is performer with the capability of both entertaining and irritating his audience, and his autobiography has much the same qualities. It contains both the outrageous and the mundane; it is both factual and opinionated. Rooney's claim that Lana Turner became pregnant by him and underwent an abortion has been denied by the actress. Similarly, Norma Shearer's secretary has denied that her employee enjoyed a sexual liaison with Rooney. The reader will probably never know what is the truth. There are certainly errors in Rooney's book, such as his naming Herbert J. Yates president of Columbia Pictures, and moving his Fryman Road home from Studio City to Sherman Oaks and back again to Studio City in the space of a few pages. There is also much that is true to be found here, particularly Rooney's diatribe on the current Hollywood establishment. Some of the quotes from reviews and listings of film appearances read more like a researcher's efforts than Rooney's own outpourings. But how can one not like an actor who loves animals as much as Rooney and calls ants, spiders, crickets, moths, and the like "little people."

Selected Takes: Film Editors on Editing by Vincent LoBrutto. $49.95/$17.95. (Praeger. Hardcover/Paperback. 262 pages. Index.)

Selected Takes consists of interviews with 21 American film editors: Harold Kress, William Reynolds, Rudi Fehr, Ralph Winters, David Bretherton, Anne V. Coates, Dede Allen, Tom Rolf, Maury Wintrobe, Jerry Greenberg, Lou Lombardi, Evan Lottman, Alan Heim, Richard Halsey, Michael Kahn, Richard Marks, Tina Hirsch,

Craig McKay, Susan E. Morse, Arthur Schmidt, and Carol Littleton. Each interview is prefaced by a short commentary and a filmography consisting of titles and dates only.

A book such as this has long been needed, and author LoBrutto is to be congratulated for the range of questions he has asked and the diversity of his subjects, from legendary Hollywood veteran William Reynolds to Woody Allen's editor Susan E. Morse. Most of the interviewees acknowledge the contributions of their directors and the interrelationship between the two crafts. Some do not, and there is a danger here of overemphasizing the contributions of editors and overreliance on their memories.

For example, Rudi Fehr discusses the making of *House of Wax* without any mention of the film's director, Andre de Toth. Fehr claims that he determined the length of shots which illustrated the 3-D technique, a statement which is patently untrue in that the director would have had to shoot sufficient footage before the editor had access to it. Further, Fehr claims the film was shot in sequential order, a statement director de Toth denies to me.

In other words, this is a great book, but don't necessarily accept as gospel everything which the editors have to say. Recommended.

Echo and Narcissus: Women's Voices in Classical Hollywood Cinema by Amy Lawrence. $37.50/$15.95. (University of California Press. Hardcover/Paperback. 222 pages. Index. Illus.)

According to its publisher, *Echo and Narcissus* "expands feminist studies of film by adding voice to the agenda." In a text which is difficult, but not impossible, for the average human being (female or male) to understand, the author discusses how women's voices are used in nine films: *Sadie Thompson, Rain, Blackmail, The Spiral Staircase, Notorious, Sorry, Wrong Number, Miss Sadie Thompson, Sunset Boulevard,* and *To Kill a Mockingbird.* She also provides a rather interesting feminist overview of the presentation of female voices in early phonograph recordings.

As a male, I am probably considered unqualified to judge a book such as this, but I must wonder at the validity of the author's commentary when she discusses the British feature *Blackmail* in a book subtitled "Women's Voices in Classical Hollywood Cinema." Incidentally, the point she makes with *Blackmail* is that Anny Ondra's voice is "dubbed" by Joan Barry, and thus " 'a woman' was constructed according to the production team's idea of what a woman should be." Presumably, the production team should either have fired Anny Ondra and replaced her on camera with Joan Barry, or filmed the production with Anny Ondra's own voice and not worried that no one in the audience would understand her.

Totally Uninhibited: The Life and Wild Times of Cher by Lawrence J. Quirk. $20. (William Morrow. Hardcover. 303 pages. Index. Illus.)

"Superstar" is a much overused expression and one that can only be used to describe three of the current crop of female performers: Barbra Streisand, Madonna, and Cher. Interestingly, all three are known throughout the world by single names—indeed two utilize only one name and the other is immediately identifiable by her first or last name—and all are equally adept as singers or actresses. Of the group, I prefer Cher whose outgoing personality is as pleasantly appealing as the variety of her screen characterizations.

It is perhaps surprising to find Lawrence J. Quirk, whose previous books have considered the lives of Margaret Sullavan, Norma Shearer, and Bette Davis, offering a biography of an entertainment figure as contemporary as Cher, but he has done an excellent job of chronicling the lady's life and career. Despite this being an unauthorized biography, the book is kindly disposed towards Cher, and I am sure there is nothing here that either the performer or her fans will find offensive. The book also includes a selective discography and bibliography, and a filmography of Cher's nine feature films from *Good Times* in 1967 through *Mermaids* in 1990.

A Terrible Liar: A Memoir by Hume Cronyn. $23. (William Morrow. Hardcover. 431 pages. Index. Illus.)

A Terrible Liar is the life story of Hume Cronyn up through 1966, documenting his work on stage from 1930, on screen from 1942, his marriage to Jessica Tandy, and much, much more. Hume Cronyn is honest and forthright in what he writes, but I regret he failed to hold my interest. Certainly I am not eagerly awaiting the second volume. Sorry!

Horror Film Stars by Michael R. Pitts. $24.95. (McFarland. Paperback. 475 pages. Index. Illus.)

This is a revised and expanded edition of the book which first appeared in 1981. *Horror Film Stars* contains career articles examining the lives and film appearances of 17 stars and 49 players. The former are Lionel Atwill, John Carradine, Lon Chaney, Sr., Lon Chaney, Jr., Peter Cushing, Boris Karloff, Christopher Lee, Peter Lorre, Bela Lugosi, Paul Naschy, Donald Pleasence, Vincent Price, Claude Rains, Basil Rathbone, Tod Slaughter, Barbara Steele, and George Zucco, and for each, author Pitts has provided a complete filmography, listing title, producer/distributor, and year of release. The other group (which includes John Agar, Yvonne De Carlo, Dwight Frye, Elsa Lanchester, Cameron Mitchell, Dennis Price, and Fay Wray) are players not exclusively associated with the horror genre, and their filmographies are limited to genre appearances.

As might be expected, Michael R. Pitts has provided an easy-to-read reference work, which is both reliable and informative. I might point out that unlike all the McFarland books which I reviewed over the past three years, *Horror Film Stars* is published in paperback, and subsequently far more affordable for individuals (who I am sure will also appreciate that it is stitch-bound and, therefore, should not fall apart too easily).

High-Class Moving Pictures: Lyman H. Howe and the Forgotten Era of Traveling Exhibition, 1880-1920 by Charles Musser, in collaboration with Carol Nelson. $45. (Princeton University Press. Hardcover. 384 pages. Index. Illus.)

Influenced by a documentary made a few years ago by Carol Nelson, *High-Class Moving Pictures* is an extremely detailed history of the life and career of traveling showman Lyman H. Howe, who operated out of Wilkes-Barre, Pennsylvania, bringing screen entertainment to those who were not within easy reach of a nickelodeon or vaudeville house in the late 1890s and early 1900s. In many respects, Howe was similar to the better known Burton Holmes.

Despite being an academic text, *High-Class Moving Pictures* is readable, although I suspect most will use it as a reference source rather than attempt to sit down and actually peruse it. (Despite being well worthwhile, it is not an exciting book.) There are a number of appendices, including a very welcome Lyman H. Howe filmography. Recommended.

Movie Love: Complete Reviews 1988-1991 by Pauline Kael. $14.95. (A William Abrahams Book/Plume. Paperback. 360 pages. Index.)

In March 1991, Pauline Kael retired as film critic of *The New Yorker,* and so this is probably the final volume of her collected reviews. Eighty-five such reviews are reprinted here, beginning with *Bird* and ending with *L. A. Story.* All make intelligent and elegant reading as one would expect from a lady who was, and as long as she lives will remain, America's finest film critic. Miss Kael is generally noted for the length of her reviews, and yet at the same time in rereading them one is struck by her use of a paucity of words to put across a statement: of Tom Cruise in *Born on the Fourth of July:* "a cute kid and his face sells magazines"; of *The Tall Guy:* "This picture is the opposite of black comedy; it's friendly"; of *Edward Scissorhands:* "Beatlejuice would have spit in this movie's eye." Note the use of the word "movie" in the last quote. Miss Kael has a satisfying nonintellectual approach to the medium—they are movies and she is a movie reviewer. Bravo!

Costume Design in the Movies: An Illustrated Guide to the Work of 157 Great Designers by Elizabeth Leese. $13.95. (Dover. Paperback. 171 pages. Illus.)

This is a welcome reprint of the 1976 reference book on costume design, with entries updated to 1988. For each designer, the author provides a brief biographical sketch, sometimes as short as one line, and a complete list of film credits. Oscar and British Academy of Film and Television Arts nominees and winners for costume design are listed in an appendix, and a film title index provides easy access to the name(s) of that production's costume designer(s). Because *Costume Design in the Movies* was originally a British publication, a somewhat simplistic introductory essay is oriented towards British films, British stars, and British designers.

The Unabridged James Dean: His Life and Legacy from A to Z by Randall Riese. $35. (Contemporary Books. Hardcover. 603 pages. Illus.)

This is an extraordinary book—a popular, encyclopedic reference book on James Dean. Its hundreds of entries include players and directors who worked with the actor, his films and television programs, cities associated with the actor, and much, much more. There are entries for Dean's shoe size, waist size, the National Westminster Bank (which used Dean in a 1987 advertising campaign), Pacific Indemnity (which had insured Dean's life at the time of his death). There is an entry discussing the question of whether Dean was homosexual, an entry on calendars which featured Dean, and even on the makeup artists who worked on his various films.

To be honest, I don't really care that much about James Dean, and still I found this a fascinating volume. Dean fans will love it, as will reference librarians answering those esoteric questions on this contemporary American folk hero. Recommended.

Apparently Incongruous Parts: The Worlds of Malcolm Lowry edited by Paul Tiessen. $27.50. (Scarecrow Press. Hardcover. 242 pages.)

The Cinema of Malcolm Lowry: A Scholarly Edition of Lowry's 'Tender Is the Night,' edited by Miguel Mota and Paul Tiessen. $34.95. (University of British Columbia Press. Hardcover. 278 pages. Index.)

These two recent books on novelist Malcolm Lowry have some interest for film scholars and film enthusiasts. *Apparently Incongruous Parts* is a collection of academic essays (readable because they are written by those outside of university film departments), including "A Canadian Film Critic in Malcolm Lowry's Cambridge" by Paul Tiessen and "The 67th Reading: *Under the Volcano* and Its Screenplays" by Wieland Schulz-Keil.

The Cinema of Malcolm Lowry begins with an extensive essay on Lowry's interest in the motion picture, his comments on various films (notably those which he viewed at the Vancouver Film Society), and his marriage to former silent screen actress Margerie Bonner. The remainder of the book is taken up with Lowry's screenplay based on F. Scott Fitzgerald's *Tender Is the Night.* Reading the screenplay, it is obvious why the Hollywood film industry had little or no interest in it. Lowry does not understand how to write a film script and was naive enough to believe it was possible to write a screenplay based on a well-known novel on a speculative basis. *The Cinema of Malcolm Lowry* is well worth reading for both its parts and reminds us what a fascinating character was Malcolm Lowry and what an explosive and amusing "fictional" film could be made of his life. Recommended.

Babel & Babylon: Spectatorship in American Silent Film by Miriam Hansen. $37.50. (Harvard University Press. Hardcover. 387 pages. Index. Illus.)

Babel & Babylon is a study of the emergence of an audience for the motion picture in relationship to the development of the medium in the silent era, with particular reference to *Intolerance* and the films of Rudolph Valentino. It is an interesting idea, but, sadly, the book is almost unreadable because of the author's insistence on using academic jargon. The following is typical of the text:

"Although scopophilia as a component drive does not exist outside the norms of genitality, collapsing it a priori into an Oedipal teleology restricts the possibility of conceptualizing the female spectator (or, for that matter, the male spectator as well) in other than binary, ahistorical terms, it thus tends to reproduce a phallic economy on the level of critique."

The author is professor of English at the University of Chicago.

Tomart's Price Guide to Radio Premium and Cereal Box Collectibles by Tom Tumbusch. $22.95. (Wallace-Homestead. Paperback. 176 pages. Illus.)

Who would have thought there were so much radio-related memorabilia and so reasonably priced (at least in comparison to movie memorabilia)? The book is arranged by radio program, with a brief comment on each and a listing of memorabilia and its value if good, fine, or mint. Some of the entries are considerable; for example, *Howdy Doody* runs three pages and includes such items as a Howdy for President Flip-Up Badge and a Flasher Ring. There is also a 23-page entry on cereal boxes (which I had no idea people collected). Did you know that a Wheaties box (presumably empty) with Tarzan on the back could be worth as much as $600?

Radio Premium and Cereal Box Collectibles is profusely illustrated, with an eight-page color section. Even if you are not into collecting, the book is fun to browse through.

The Comics Come Alive: A Guide to Comic-Strip Characters in Live-Action Productions by Roy Kinnard. $32.50. (Scarecrow Press. Hardcover. 252 pages. Index. Illus.)

The Comics Come Alive is a useful reference guide to comic strip characters on screen and television, from "Ace Drummond" to "Wonder Woman." The comic characters' original incarnations as newspaper strips or comic books are noted, and the author then provides detailed credits for big and small screen appearances, together with a short essay on the character. A Miscellaneous Entries section covers such television items as *The Fabulous Funnies* and *Legends of the Superheroes.*

The Comics Come Alive is up-to-date enough to include *Teenage Mutant Ninja Turtles.* Curiously, while the author dates the first appearances of his other characters, he notes only that the Turtles first appeared as an underground comic book of unidentified vintage.

Cruel City: The Dark Side of Hollywood's Rich & Famous by Marianne Ruuth. $19.95. (Roundtable Publishing. Hardcover. 204 pages. Illus.)

Cruel City is a rehash of the same old Hollywood tragedies and scandals—F. W. Murnau, William Desmond Taylor, Marilyn Monroe, etc.—arranged by chapter headings covering the manner of the participants' deaths—mysterious, suicide, murder, etc. It is all rather tedious and lacking any new facts or thoughts. It is hardly, as the publisher claims in the dust jacket blurb, "the first book of its kind to reveal the control once exerted by the industry."

American Film and Society since 1945 by Leonard Quart and Albert Auster. $12.95. (Praeger. Paperback. 204 pages. Index.)

This is a second edition of a book first published in 1984 and is revised and updated by Leonard Quart. Decade by decade, the authors provide a highly readable examination of American history and its interrelationship to the popular and critically acclaimed Hollywood feature films of the period. I suppose that this book might be considered old-fashioned by many in the academic community in that the comments by the authors are both understandable and informative. It is for that reason that I admire the text and can understand how helpful a book such as this must be in the classroom. Recommended.

The Cult Film Experience: Beyond All Reason edited by J. P. Telotte. $36/$15.95. (University of Texas Press. Hardcover/Paperback. 225 pages. Index. Illus.)

J. P. Telotte has gathered together a collection of essays on cult films and personalities, with *Casablanca* receiving the most attention. The genre of the cult film is subdivided into older Hollywood movies and more recent "midnight movies," of which the most obvious example is *Rocky Horror Picture Show.* The academics invited to contribute to this book do not on the whole have too much original to say, but at least what they have to say, or rather write, is understandable. And I suppose having academics discuss *Rocky Horror Picture Show* or the work of Edward Wood, Jr. gives respectability to these subjects.

Comedy/Cinema/Theory edited by Andrew S. Horton. $35/$12.95. University of California Press. Hardcover/Paperback. 258 pages. Index. Illus.)

Academics do it to film comedy—and do they ever. *Comedy/Cinema/Theory* is a collection of essays on quite a variety of comic-related subjects, including the sight gag, Charlie Chaplin, *Mrs. and Mrs. Smith,* Frank Tashlin, Preston Sturges, Jerry Lewis, the Three Stooges, and *Zelig.* Of course, the writers do just about everything they can to destroy the comedic propensities of the subjects about which they write. For example, we learn that Woody Allen's line, "I believe there is an intelligent spirit that controls the universe, except certain parts of New Jersey," demonstrates the process of biosociation in practice, "the sudden flash of recognition as 'A' (the theological question) and 'B' ('certain parts of New Jersey') collide."

However, there is one brilliant essay in *Comedy/Cinema/Theory,* whose relationship to screen comedy I am a little uncertain of, and that is on the Hollywood film industry's ability to film the male nude form without showing genitalia, as exampled in *Once Bitten, Bull Durham, The Last Temptation of Christ,* and other productions. Apparently, the author of this piece, one Peter Lehman, has written a forthcoming book on sexuality and the representation of the male body. I cannot wait.

December 1991

Leonard Maltin's Movie and Video Guide 1992 by Leonard Maltin. $7.99. (A Signet Book. Paperback. 1,502 pages.)

The new edition of Leonard Maltin's *Movie and Video Guide* is promoted as an "annual # 1 Best Seller." It is also the No. 1 reference book of any year, with its now familiar format of taut, sometimes amusing two- or three-line critiques, with basic credits, for some 19,000 feature films. There is also indication as to which films are also available on home video, an extremely valuable listing of mail-order sources for video, and title indices for major stars and directors. There is little to be written which has not already been said about this annual compendium, except that at $7.99 it is undoubtedly the most reasonably priced film book on the market today and as such an ideal Christmas gift for anyone, even yourself. Highly Recommended.

Madam Valentino: The Many Lives of Natacha Rambova by Michael Morris. (Abbeville Press. Hardcover. 272 pages. Index.)

Natacha Rambova is best remembered (if she is remembered at all) as the second wife of Rudolph Valentino. However, as this book's subtitle indicates, she had many other claims to fame: as a protegé of Elsie de Wolfe, the mistress of Theodore Kosloff, a participant in the Spanish Civil War, a couturiere, and a highly regarded Egyptologist. All of these activities are discussed in detail in this beautifully produced and illustrated volume, which is a credit to both its author and publisher.

Natacha Rambova led an extraordinary life, far removed from her beginnings as Winifred Shaughnessy, the daughter of a Federal Marshall in Utah. Readers will be fascinated by the author's account of her life with Valentino and her efforts to shape his career. As am I, they will be grateful to the author for denying many of the so-called "facts" about Rambova, including her supposed lesbianism, and for his demolishing of some outrageous statements made by Kenneth Anger in *Hollywood Babylon.* Highly Recommended.

Images of Australia: 100 Films of the New Australian Cinema by Neil Rattigan. $24.95/$12.95. (Southern Methodist University Press. Hardcover/Paperback. 347 pages. Illus.)

Images of Australia is a curious book to appear from Southern Methodist University Press, but that does not mean that it is not a welcome one. Here is a major reference work on Australian cinema from the 1970s through the present. For 100 feature films, the author provides full credits (including the U. S. title if different from the original), a synopsis and a readable comment. All the popular films, such as *Crocodile Dundee* and *Gallipoli,* are here as well as many lesser known titles. It should not be necessary to inform anyone interested in the motion picture of the importance of Australian filmmaking of the past two decades, but if there is any doubt as to the quality and variety of Australian feature films, *Images of Australia* puts it to rest.

The author has also included a glossary of Australian terms, and it makes fascinating reading. I learned, for example, that to "spifflicate the poms" is to beat the English at cricket. Of course, that is something Australians seldom do. Recommended.

British Cinema: The Lights That Failed by James Park. $34.95. (Trafalgar Square. Paperback. 192 pages. Index. Illus.)

I cannot agree with the publisher's description of this book as a "controversial history of British filmmaking," but it is a first-rate concise history of the subject, with just enough opinion to make it entertaining. (The only two controversial figures I can think of in

British films are Derek Jarman and Peter Greenaway, both of whom receive two mentions apiece here.) All the important productions and industry figures are discussed, albeit briefly, and the author includes a brief bibliography. This is a perfect textbook for any college course on British cinema and a pleasant read for anyone with little more than a passing interest in the subject. Recommended.

Clint Eastwood/Malpaso by Fuensanta Plaza. $24.95. (Ex Libris. Paperback. 246 pages. Index. Illus.)

This is a book-length study of the feature films produced by Clint Eastwood's company Malpaso, all starring the actor, beginning with *Hang 'Em High* (1968) and concluding with *White Hunter, Black Heart* (1990). For each film, complete credits are provided, together with brief comments by Eastwood and an essay by the author. The book also includes a collection of comments by Eastwood on filmmaking and a list of actors recurring in Malpaso films.

This is a "Films of" book which is not in the usual "Films of" mould. First, it is a beautiful book to hold and experience. There is a sense of taste and a style to the layout and design, enhanced by the color photographs, arranged one for each film. Secondly, the text is not written in film buff style, but adopts an old-fashioned elegance with simple phrases utilizing well-chosen words.

The author is quite obviously in love with his subject and is not about to find fault with Mr. Eastwood or his films. This is not a critical work. It is a book which every Eastwood fan will want to own, and one which would not be out of place on the bookshelf of any discerning bibliophile. Recommended.

Notes on the Making of Apocalypse Now by Eleanor Coppola. $14.95. (Limelight Editions. Paperback. 292 pages. Illus.)

This is the paperback reprint of the 1979 book in which Eleanor Coppola documents the making of her husband's controversial feature film *Apocalypse Now,* together with the declining fortunes of her marriage. The reprint also includes a new afterword bringing the story up-to-date, revealing, to those unaware, that the Coppola marriage did survive and that the documentary which Eleanor Coppola planned to make was completed by others for screening on the Showtime cable channel in the fall of 1991. It is odd that Mrs. Coppola does not bother to credit the filmmakers who took her unedited footage and interviews and made the documentary reality.

As with all diaries, journals, and notes written with publication as an ultimate end, there is a pretention here which is irritating to all readers not totally enamoured of Coppola's work. Neither he nor his wife come across as particularly lovable human beings, and Coppola's lack of foresight and planning in the production of *Apocalypse Now* is staggering. Does this guy know how to make films?

Loser Take All: The Comic Art of Woody Allen by Maurice Yacowar. $14.95. (Continuum. Paperback. 330 pages. Index. Illus.)

This is an expanded edition of a book which first appeared in 1979. In the intervening years, a number of volumes on Woody Allen have appeared and the importance of this work has obviously diminished. Nonetheless, *Loser Take All* is a worthwhile and highly readable critical examination of Allen's films and theatre work, with a number of general chapters together with chapter-length studies of Allen's individual achievements. Based on the annotations alone, it is obvious that the author has read everything on his subject, but, of course, was not given the opportunity to talk directly with Allen. Also included is a detailed filmography and a lengthy bibliography.

Irene Dunne: A Bio-Bibliography by Margie Schultz. $45. (Greenwood Press. Hardcover. 328 pages. Index. Illus.)

Margie Schultz was responsible for an excellent bio-bibliography of Ann Sothern. She has now turned her attention to Irene Dunne, and accomplished a similarly superb job, No. 19 in Greenwood's Bio-Bibliographies in the Performing Arts series.

The book begins with an effusive but accurate biography. It continues with a listing of Miss Dunne's plays and personal appearances, a filmography, listings of radio and television work, awards and honors, an annotated bibliography, and concludes with a selected listing of magazine covers graced by Miss Dunne. Yet again, I was impressed by the exhaustive detail of Margie Schultz's research. For example, her filmography includes complete credits, lengthy synopses, a sampling of reviews, and relevant notes. Recommended.

Perspectives on Chinese Cinema edited by Chris Berry. $51/$21.95. (Indiana University Press. Hardcover/Paperback. 241 pages. Index. Illus.)

A second edition of a book first published in 1985, *Perspectives on Chinese Cinema* is a valuable collection of essays on the development of contemporary Chinese filmmaking, with chapters also covering the cinema in Taiwan and Hong Kong. As important as the essays themselves are the appendices, with biographies of major Chinese directors, a glossary of Chinese characters, and a chronology of both Chinese general history and cinema history from 1894 onwards. As the editor points out in his introduction, *Perspectives on Chinese Cinema* was conceived prior to the Tiananmen Square massacre, and it remains to be seen how Chinese filmmaking will be affected as a result of the crackdown. Recommended.

Playing with Power: In Movies, Television and Video Games by Marsha Kinder. $22.50. (University of California Press. Hardcover. 277 pages. Index. Illus.)

In *Playing with Power,* a film professor from the University of Southern California examines how children's minds are influenced by what they see on television, in the movies, and in video games. She argues the obvious, that such propagandistic programming urges children to buy, or at least persuade their parents to buy, specific products.

The author's writing is, apparently, based on recent work in feminist psychoanalysis, Althusser's economic theories of social indoctrination, Piaget-inspired studies of child development, and studying the behavior of her own small son. With such a research background, it should be obvious that *Playing with Power* is neither for you nor me. In all honesty, I found the most interesting section of the book to be the Preface, from which I learned that the author's editor at the University of California Press is a "longtime friend and collaborator" (one way to ensure publication), that there is an Institute for the Study of Women and Men (why not Men and Women?), and that one of the author's friends has accused her of writing "on texts that are 'unworthy' of my powers of analysis." With friends like that to massage one's ego, who needs a good review?

Mary Martin: A Bio-Bibliography by Barry Rivadue. $39.95. (Greenwood Press. Hardcover. 249 pages. Index. Illus.)

Mr. Rivadue has done a superlative job in documenting the career of Mary Martin. The book opens with a 22-page biography, which is perhaps a little uncritical, but which wins my praise for its author's continued reference to his subject by her surname rather than the effusive "Mary." The list of stage appearances which follows is detailed in the extreme, including sample reviews, complete credits, listings of musical numbers, and the author's general comments. Somewhat less detail is provided for Miss Martin's film work, which must, of course, be considered of secondary importance, but for which Mr. Rivadue still provides more than adequate documentation. Additional listings cover Miss Martin's work on radio, television, and in the recording industry. There is also, of course, an annotated bibliography, plus a list of awards, a useful chronology, and a listing of products endorsed by the actress from the 1930s through the 1980s.

My only criticism is that the section documenting archival sources could have been lengthened with the addition of non-New York institutions: the address for the Academy's Center for Motion Picture Study is not complete and incorrect as it appears, and if one commercial photo source (Photofest) is included then others (such as Eddie Brandt's Saturday Matinee) should also be included. This is, however, a minor quibble regarding a section not usually included at all in other Greenwood Bio-Bibliographies in the Performing Arts, of which Mary Martin is No. 18. Recommended.

Jessica Tandy: A Bio-Bibliography by Milly S. Barranger. $39.95. (Greenwood Press. Hardcover. 158 pages. Index. Illus.)

Yet another book—No. 22—in the Bio-Bibliographies in the Performing Arts series. This one is prepared by a professor at the University of North Carolina and I won't say that it is better written than some of the earlier volumes but it does boast a more formal text. In all other respects, it is similar to the other volumes, and like the Mary Martin book concentrates, quite naturally, on the subject's stage career. Also included is a chronology, a filmography, a list of television appearances, a discography, a listing of awards and nominations, and the usual annotated bibliography. I was interested to learn that Jessica Tandy made her film debut in 1932, in the British feature *The Indiscretions of Eve,* playing the part of the maid. Recommended.

Reel Women: Pioneers of the Cinema, 1896 to the Present by Ally Acker. $34.95. (Continuum. Hardcover. 400 pages. Index. Illus.)

Reel Women is a useful reference work on women in the film industry (both American and foreign), active behind the camera from the 1890s to the present. The book is divided into sections on directors, producers, editors, etc., and specific women are discussed in each category. Useful as is this book, it is neither as revolutionary not as scholarly as some of the overenthusiastic reviews from other publications might indicate. The author appears to have made little attempt to view most of the films under discussion and to have conducted little original research, preferring to rely on secondary sources such as my own *Early Women Directors.*

In return for the use of the many fascinating illustrations in the volume, the author was also, presumably, required to allow the owner of the photographs, Marc Wanamaker, a two-page "Afterword." Here, Wanamaker makes the extraordinary statement that Natalie Kalmus is "forgotten," whereas, of course, she is remembered because her name appears on virtually every Technicolor film ever made and ever screened. And why mention Kalmus, who it is very apparent contributed nothing to most of the films on which her name appears?

Robert Altman's America by Helene Keyssar. $16.95. (Oxford University Press. Paperback. 385 pages. Index. Illus)

Robert Altman's America is a scholarly study of the films of Robert Altman (*M*A*S*H, Nashville, A Wedding,* etc.), grouped under a variety of categories. For me, the most interesting section of the book was its preface, in which the author explains the book's origins and growth within her own life and career development. *Robert Altman's America* is understandable and informative, but will doubtless serve primarily as a textbook rather than a volume for the casual reader.

From Limelight to Satellite: A Scottish Film Book edited by Eddie Dick. $64.95/$23.95. (Indiana University Press. Hardcover/Paperback. 256 pages. Illus.)

The complete gamut of filmmaking in Scotland is covered in this new book, first published in the United Kingdom by the British Film Institute. A series of well-written and intelligent essays cover such topics as Scottish silent films, documentary production in Scotland, the Edinburgh International Film Festival, *Whiskey Galore* (known in the U. S. as *Tight Little Island*), directors Bill Douglas and Bill Forsythe, and Sean Connery. Also included is a filmography of Scottish-related films from 1898 through 1990. The filmography is indexed but, regrettably, the rest of the volume is not.

This is an extremely important reference source on cinema in a part of Europe better known for its scenery and the friendly nature of its people than its filmmaking efforts. In his current attempts to promote independence for Scotland, Sean Connery may create a vital film industry there. Bill Douglas and Bill Forsythe (who made the brilliant *Local Hero* and *Comfort and Joy*) have clearly demonstrated that Scotland is capable of producing some unique films. Highly Recommended.

Tallulah Bankhead: A Bio-Bibliography by Jeffrey L. Carrier. $42.95. (Greenwood Press. Hardcover. 286 pages. Index. Illus.)

New books in the Bio-Bibliographies in the Performing Arts series from Greenwood Press seem to be arriving almost daily in the mail (or to be more precise via the UPS van). The latest, No. 21, is devoted to Tallulah Bankhead, and like other recent additions to the series is a superior effort. It begins with a preface, capturing the essence of Bankhead's personality, and is followed by a well-considered, 44-page biographical sketch. In great detail, the author documents the work of the actress on stage, screen, radio, and television. Also included is the obligatory annotated bibliography; a listing of awards, honors, and tributes; a listing of film roles considered and film tests; and a listing of books and films which have utilized the Bankhead name. Included in the last is one of my favorite mystery novels, *The Tallulah Bankhead Murder Case* by George Baxt, which may contain factual inaccuracies (as author Carrier points out), but is still great fun.

Each new book in this series seems better than the last, and in the case of the current offering even looks better, in that this volume is typeset rather than merely typed. Because it is a serious study of its subject, the book lacks some of the more outrageous Bankhead humor, but there is sufficient here to delight the fan and more than enough valuable resource material to please the scholar and student. Recommended.

List of Publishers

Abbeville Press, Inc., 488 Madison Avenue, New York, N. Y. 10022
Harry N. Abrams, Inc., 100 Fifth Avenue, New York, N. Y. 10011
ALA Books, 50 East Huron Street, Chicago, Il. 60611
W. H. Allen, Sekforde House, 175-179 St. John Street, London, United Kingdom EC1V 4LL
Archival Photography, 14845 Anne Street, Allen Park, Mi. 48101
Artist's Press, P. O. Box 22482, Minneapolis, Mn. 55422
ACS Press, P. O. Box 2230, Hollywood, Ca. 90078
Antheneum, 866 Third Avenue, New York, N. Y. 10022

Baker Street Publications, P. O. Box 994, Metairie, La. 70004
Ballantine Books, 201 East 50th Street, New York, N. Y. 10022
Bantam Books, 666 Fifth Avenue, New York, N. Y. 10103
The Bare Facts, P. O. Box 3255, Santa Clara, Ca. 95055
Birch Lane Press, 600 Madison Avenue, New York, N. Y. 10022
Earl Blair Enterprises, P. O. Box 87, Toney, Al. 35773
British Film Institute, 21 Stephen Street, London W1P 1PL, United Kingdom
Broadview Press, 421 Center Street, Lewiston, N. Y., 14092
Broadway Press, 12 West Thomas Street, Shelter Island, N. Y. 11964
Business One Irwin, 1818 Ridge Road, Homewood, Il. 60430

Cambridge University Press, 40 West 20 Street, New York, N. Y. 10011
Campfire Video, P. O. Box 71318, Los Angeles, Ca. 90071-0318
CCM Associates, Studio A, Sea View Drive, El Cerrito, Ca. 94530
La Cineteca del Friuli, via Osoppo 26, 1-33014 Gemona del Friuli (UD), Italy
Citadel Press, 600 Madison Avenue, New York, N. Y. 10022
Columbus Books, 19-23 Ludgate Hill, London EC4M 7PD, United Kingdom
Contemporary Books, Inc., 180 North Michigan Avenue, Chicago, Il. 60601
Continuum, 370 Lexington Avenue, New York, N. Y. 10017
Creatures at Large Press, P. O. Box 687, Pacifica, Ca. 94044
The Crossing Press, Freedom, Ca. 95019

Crown Publishing Group, 201 East 50 Street, New York, N. Y. 10022

Da Capo Press, 233 Spring Street, New York, N. Y. 10013-1578
Denlinger's Publishers, Box 76, Fairfax, Va. 22030
Doubleday, 666 Fifth Avenue, New York, N. Y. 10103
Dover Publications, Inc., 180 Varick Street, New York, N. Y. 10014
E. P. Dutton, 1633 Broadway, New York, N. Y. 10019

Empire Publishing, Route 3, Box 83, Highway 220 South, Madison, N. C. 27025
Ex Libris, 540 Santa Cruz Avenue, Suite 117, Los Gatos, Ca. 95030

Facts on File, Inc., 400 Park Avenue South, New York, N. Y. 10016
Fairleigh Dickinson University Press, 285 Madison Avenue, Madison, N. J. 07940
Farrar, Straus & Giroux, Inc., 19 Union Square West, New York, N. Y. 10003
Donald I. Fine, Inc. 19 West 21 Street, New York, N. Y. 10010
Focal Press, 80 Montvale Avenue, Stoneham, Ma. 02180
Samuel French, Inc., 7623 Sunset Boulevard, Hollywood, Ca. 90046

Garland Publishing, Inc., 136 Madison Avenue, New York, N. Y. 10016
Le Giornate del Cinema Muto, c/o La Cineteca del Friuli, via Osoppo 26, 1-33013 Gemona (UD), Italy.
Gower Publishing Company, Old Post Road, Brookfield, Vt. 05036
Greenwood Press, Inc., Box 5007, Westport, Ct. 06881
G. K. Hall & Co., 70 Lincoln Street, Boston, Ma. 02111

Harper & Row, Inc., 10 East 53 Street, New York, N. Y. 10022
Harper Collins, 10 East 53 Street, New York, N. Y. 10022
Harvard University Press, 79 Garden Street, Cambridge, Ma. 02138-9983
Heinemann Educational Books, Inc., 361 Hanover Street, Portsmouth, N. H. 03801-3959
Hippocrene Brooks, Inc., 171 Madison Avenue, New York, N. Y. 10016
The Holland House, 17142 Index Street, Granada Hills, Ca. 91344

Henry Holt & Co., 115 West 18 Street, New York, N. Y. 10011

Indiana University Press, 10 and Morton Streets, Bloomington, In. 47405

Institute for Cinema and Culture, Communication Studies Building, University of Iowa, Iowa City, Ia. 52242

The Johns Hopkins University Press, Baltimore, Md. 21211

Alfred A. Knopf, Inc., 201 East 50 Street, New York, N. Y. 10022

LA 411 Publishing, P. O. Box 480495, Los Angeles, Ca. 90048

Laughs Unlimited Incorporated, 1124 Firehouse Alley, Sacramento, Ca. 95814

Lehigh University Press, 440 Forsgate Drive, Cranbury, N. J. 08512

Limelight Editions, 118 East 30 Street, New York, N. Y. 10016

Little, Brown & Co., Inc., 34 Beacon Street, Boston, Ma. 02108

Madison Books, 4720 Boston Way, Lanham, Md. 20706

MagicImage Filmbooks, 740 S. 6 Avenue, Absecon, N. J. 08201

McFarland & Company, Inc., Box 611, Jefferson, N. C. 28640

McPherson and Company, P. O. Box 1126, Kingston, N. Y. 12401

Gerald McKee, 34 Ashlea Road, Gerrards Cross, Bucks SL9 8NY, United Kingdom

Mercury House, Suite 400, 201 Filbert Street, San Francisco, Ca. 94133

Milestone, P. O. Box 2748, Laurel, Md. 20708

Mission Publishing, P. O. Box 1440, Santa Barbara, Ca. 93121

Moonstone Press, P. O. Box 142, Beverly Hills, Ca. 90213

William Morrow & Co., Inc., 105 Madison Avenue, New York, N. Y. 10016

Museum of Modern Art, 11 West 53 Street, New York, N. Y. 10019

National Center for Film and Video Preservation, 2021 North Western Avenue, Los Angeles, Ca. 90027

Neville Publishers, Inc., P. O. Box 5056, Santa Barbara, Ca. 93150

New American Library, 375 Hudson Street, New York, N. Y. 10014

W. W. Norton & Co., Inc., 500 Fifth Avenue, New York, N. Y. 10110

Old Mill Books, Box 100, Eisenhart Mill Road, East Berlin, Pa. 17316
O'Raghailligh Ltd., 740 South 6 Avenue, Abescon, N. J. 08201
The Overlook Press, 12 West 21 Street, New York, N. Y. 10010
Oxford University Press, Inc., 200 Madison Avenue, New York, N. Y. 10016

Pantheon Books, Inc., 201 East 50 Street, New York, N. Y. 10022
Penguin Books, 375 Hudson Street, New York, N. Y. 10014
Perigree Books, 200 Madison Avenue, New York, N. Y. 10016
Pineapple Press, P. O. Drawer 16008, Sarasota, Fl. 34239
Plume/ E. P. Dutton, 375 Hudson Street, New York, N. Y. 10014
PM Publications, P. O. Box 180182, Austin, Tx. 78718-0182
Praeger, One Madison Avenue, New York, N. Y. 10010
Prelinger Associates, Inc., 430 West 14 Street, New York, N. Y. 10014
Prentice-Hall, see Simon & Schuster
Princeton University Press, Princeton, N. J. 08540
Prometheus Books, 700 East Amherst Street, Buffalo, N. Y. 14215
Publications International Ltd., 7373 North Cicero Avenue, Lincolnwood, Il. 60646

Quality R. J. Press, 848 North Dillon Street, Los Angeles, Ca., 90026
Quinlan Press, 131 Beverly Street, Boston, Ma. 02114

Random House, Inc., 201 East 50 Street, New York, N. Y. 10022
Rastar Press, c/o Australia in Print, 3350 West Bayshore Road, Palo Alto, Ca. 94303-0986
Roundtable Publishing, Inc., 29169 Heathercliff Road, Malibu, Ca. 90265
Routledge, Chapman & Hall, 29 West 35 Street, New York, N. Y. 10001
Rutgers University Press, 109 Church Street, New Brunswick, N. J. 08901

St. Martin's Press, Inc., 175 Fifth Avenue, New York, N. Y. 10010
Santa Teresa Press, 1980 Cliff Drive, Suite 224, Santa Barbara, Ca. 93109
Scarborough House, Box 459, Chelsea, Mi. 48118

Scarecrow Press, Inc., P. O. Box 4167, Metuchen, N. J. 08840
Schiffer Publishing Ltd., 1469 Morstein Road, West Chester, Pa. 19380
Charles Scribner's Sons, 866 Third Avenue, New York, N. Y. 10022
Sea-Maid Press, P. O. Box 1322, Newport Beach, Ca. 92659
Searle's Autographs, P. O. Box 849, Woodbine, Ga. 31569
Seven Arts Press, P. O. Box 649, Hollywood, Ca. 90078-0649
Shapolsky Publishers, Inc., 136 West 22 Street, New York, N. Y. 10011
Signet Books, 375 Hudson Street, New York, N. Y. 10014
Silman-James Press/Samuel French, 7623 Sunset Boulevard, Hollywood, Ca. 90046
Simon & Schuster, 1230 Avenue of the Americas, New York, N. Y. 10020
Southern Illinois University Press, P. O. Box 3697, Carbondale, Il. 62902-3696
Southern Methodist University Press, Box 415, Dallas, Tx. 75275
Stewart, Tabori & Chang, 708 Broadway, New York, N. Y. 10003
Strawberry Hill Press, 3848 S. E. Division Street, Portland, Or. 97202
Lyle Stuart, 600 Madison Avenue, New York, N. Y. 10022
The Swedish Film Institute, Box 27126, 10252 Stockholm, Sweden
Syracuse University Press, 1600 Jamesville Avenue, Syracuse, N. Y. 13244-5160

Tale Weaver Publishing, 636 North Robertson Boulevard, West Hollywood, Ca. 90069
Lance Tapley, 86 Winthrop Street, Augusta, Me. 04330
H. Arthur Taussig, 2404 Narbonne Way, Costa Mesa, Ca. 92627-1424
Temple University Press, Broad and Oxford Streets, Philadelphia, Pa. 19122
te Neues, 15 East 76 Street, New York, N. Y. 10021
Trafalgar Square, North Pomfret, Vt. 05053

University of British Columbia Press, 6344 Memorial Road, Vancouver, B. C., Canada V6T 1W5
University of California Press, 2120 Berkeley Street, Berkeley, Ca. 94720

University of Illinois Press, 54 East Gregory Drive, Champaign, Il. 61820
University of Minnesota Press, 2037 University Avenue S. E., Minneapolis, Mn. 55414
University of Oklahoma Press, 1005 Asp Avenue, Norman, Ok. 73019
University of Texas Press, P. O. Box 7819, Austin, Tx. 78713
University of Wisconsin Press, 114 North Murray Street, Madison, Wi. 53715-1199
University Press of Kentucky, Lexington, Ky. 40506-0024
Unwin Hyman, 8 Winchester Place, Winchester, Ma. 01890

Vestal Press Ltd., P. O. Box 97, Vestal, N. Y. 13850
Viking Press, 1633 Broadway, New York, N. Y. 10019
Villard Books, 201 East 50 Street, New York, N. Y. 10022
Virgin Books, 26 Grand Union Centre, 338 Ladbroke Grove, London W10 5AH, United Kingdom

Wadsworth Publishing Company, 10 Davis Drive, Belmont, Ca. 94002
Wallace-Homestead Book Company, One Chilton Way, Radnor, Pa. 19089-0230
Warner Books, Inc., 666 Fifth Avenue, New York, N. Y. 10103
Wayne State University Press, The Leonard N. Simons Building, 5959 Woodward Avenue, Detroit, Mi. 48202
Wisdom for Youth, 8405 Liberty Road, Baltimore, Md. 21207
The World of Yesterday, Route 3, Box 263-H, Waynesville, N. C. 28786

Author Index

Title Index

About the Author

ANTHONY SLIDE has authored or edited some fifty books on the history of popular entertainment, leading to his being dubbed "a one-man publishing phenomenon" by the Los Angeles *Times.* He is also editor of the "Filmmakers" series, published by Scarecrow Press. Among Slide's best-known works are *Early American Cinema, The Films of D. W. Griffith* (coauthored with Edward Wagenknecht), *Early Women Directors, The Vaudevillians, The American Film Industry: A Historical Dictionary* (named Outstanding Reference Source of the Year by the American Library Association), *The International Film Industry: A Historical Dictionary* (named Outstanding Academic Book of the Year by *Choice* magazine), and *Nitrate Won't Wait: A History of Film Preservation in the United States.*

Slide has served as associate archivist of the American Film Institute and resident film historian of the Academy of Motion Picture Arts and Sciences, produced a number of documentary films on silent screen personalities, and lectured widely on film history. In 1990, he was awarded an Honorary Doctorate of Letters by Bowling Green University, at which time Lillian Gish hailed him as "our preeminent historian of the silent film era."